GEOGRAPHIC PERSPECTIVES
IRAQ

To the brave men and women, both Iraqi and non-Iraqi,
who have suffered and sacrificed for freedom

Courage is the noblest of all attainments

GEOGRAPHIC PERSPECTIVES
IRAQ

Jon C. Malinowski

UNITED STATES MILITARY ACADEMY AT WEST POINT

The **McGraw·Hill** Companies

Book Team

Vice-President & Publisher   Jeffrey L. Hahn
Managing Editor   Theodore O. Knight
Director of Production   Brenda S. Filley
Developmental Editor   Ava Suntoke
Designer   Charlie Vitelli
Graphics   Michael Campbell
Typesetting Supervisor   Juliana Arbo
Proofreader   Julie Marsh
Cartography   Carto-Graphics

McGraw-Hill/Dushkin

A Division of The McGraw-Hill Companies

Cover: Copyright © 2004 Getty Images
Cover Design: Nancy Norton

Malinowski, Jon C.
Geographic Perspectives: Iraq/Jon C. Malinowski
New York: McGraw-Hill/Dushkin, © 2004.
p. ; cm.
I. Iraq—Geography. 1. Series.
915.67
0-07-294010-7

Preface

In the wake of the 2003 military intervention in Iraq, it is clear that Iraq will continue to be at the center of world attention for some time. As geographers and educators, this simple fact is a teaching opportunity that we simply could not pass up. In the wake of the attacks of 9/11, Gene Palka conceived this series as an opportunity to educate Americans about the geography of countries that were being featured nightly on every media outlet but that most people knew little about. Companions in the series are geographies of Afghanistan and North Korea. In the case of Iraq, the physical and human landscape display complex patterns that will challenge whatever government emerges from the ashes of the Hussein regime. Yet, despite Iraq's importance in current events, we noticed numerous oversimplifications among students, in the media, and in the popular conception, such as the notion that Iraq is all desert or the population 100 percent Muslim.

Iraq's diversity lends itself well to the geographer's approach. Geographers have traditionally focused their interests on both the physical environment and human patterns, and we do so here as well. As the reader will discover, physical patterns are often closely related to human patterns and knowledge of one greatly supports knowledge of the other. The Kurds of Northern Iraq, for example, cannot be fully understood without an appreciation of the rugged terrain that dominates "Kurdistan" and their history of tribal rivalry.

Therefore, we have taken a classic geographer's approach and crafted a regional geography of Iraq. The early chapters of the book outline Iraq's physical patterns, including landforms, climate, and vegetation. We then transition to chapters on historical, cultural, and political geography before discussing urban, population, economic, and medical patterns. This will afford a student or general reader the opportunity to experience the breadth of the country's geography applied in a real world context in much greater detail than would be allowed by an introductory text or short article.

But adopting a classic approach does not mean we have rejected the present. Every effort has been made to include the most up-to-date information available and to complement the written text with graphics. The dozens of maps, figures, satellite images, and tables not only enhance the text; they also represent our best effort to provide the reader with accurate and current information, culled from a variety of authoritative sources. Many of the maps and figures have been created especially for this edition to emphasize the geographic characteristics important to the authors. It is our hope that readers will spend time with the maps and figures as much as with the text to better appreciate the complexities of Iraq's human and physical patterns.

We have also made the conscious decision not to overwhelm the reader with in-depth analysis, forecasting of the future, or political debates. Many other books will surely fit that bill. The chapters here are concise, yet comprehensive, and, although written by professional geographers, do not inundate readers with excessive jargon. Rather, we have chosen to provide a solid foundation for pursuing further study of Iraq, if a reader so chooses, by introducing important issues and by providing a lengthy bibliography of scholarly and popular sources for additional reading.

We also hope that this book will find readers beyond the discipline of geography. Surely, students of political science, international relations, history, and Middle Eastern studies will benefit from the unique focus that geographers offer as a supplement to works from their own disciplines. And, of course, we feel strongly that this book offers the general public an authoritative primer for better understanding the morning newspaper and nightly newscast.

Jon C. Malinowski

Contents

1

Introduction

Wendell C. King
Eugene J. Palka

Key Points

- After September 11, 2001, Saddam Hussein's regime in Iraq once again came into the global spotlight
- A geographic perspective emphasizing regional patterns can provide a useful framework for understanding Iraq

The events of September 11, 2001, once again thrust Iraq onto the world stage. Although Saddam Hussein's regime was not directly involved in the attacks on the United States, his past aggressions, defiance of United Nations (UN) resolutions, and support of terrorist groups were threats to the international community. After toppling the Taliban in Afghanistan, U.S. attention turned to Iraq, precipitating a lengthy debate in the UN about Iraq's compliance with past agreements to disarm. Unable to reach a consensus, the United States and Great Britain, with support from other countries, entered Iraq in March 2003 to oust Saddam from his despotic rule of over a quarter century.

But despite years of news coverage and global attention, few people in the West know much about the physical or human landscape of Iraq, and many think that the country is just a desert, or that Iraqis are all Muslim Arabs. In reality, the people and landscape of Iraq display a great diversity from place to place. These geographic differences are the focus of this book.

But why try to understand Iraq? The answers depend on one's perspective. For some, Saddam Hussein's Iraq was seen as an enemy. The first rule of war, found in the earliest records of strategic studies, is, "Know the enemy, know yourself, and your victory will never be endangered. Know the ground, know the weather, and your victory will then be total." So stated Sun Tsu in *The Art of War*, written sometime before 200 B.C. and presented here in a 1963 translation. Surely a better understanding of Iraq's geography furthers our understanding of how Hussein and others manipulated ethnic identity, protected or exploited natural assets, and interacted with other countries in the region.

Others see Iraq as a war-torn country, where the United States has a major role in the difficult task of rebuilding. This undoubtedly calls for a sophisticated knowledge of Iraq's human and physical potential. For example, without a detailed understanding of religious and ethnolinguistic patterns, it will be difficult to assemble a stable representative government. Rebuilding schools and other social institutions will also require insight into local culture and regional history. Likewise, strengthening the economy will demand an analysis of the geographic strengths and weaknesses of a land faced with water shortages, vast differences in soil quality, and a forbidding climate.

Another perspective on Iraq is that of a country very much at the center of human history. Modern Iraq occupies the ancient region known as Mesopotamia. Here, thousands of years ago, many components of modern civilization were established by the Sumerians, Babylonians, and others. Later, Persian and Muslim influences added important layers to Iraq's cultural landscape. For example, Iraq is home to several of the holiest shrines of Shiite Muslims. Numerous ancient archeological sites have yet to be excavated. In many ways, to understand Iraq's history is to understand human history, with glorious highs mingling with brutal lows.

THE GEOGRAPHER'S PERSPECTIVE

Whether Iraq presents itself as a security challenge, an opportunity for future development, or a region of historical significance, a study of its human and physical landscape are considered to be geographic imperatives by the authors of this book. The authors are uniquely qualified to offer a special perspective; they are trained geographers representing the breadth of the subdisciplines of geography within both the physical and human branches. In addition, they are dedicated to furthering a more complete understanding of our world. The tools and methods of the geographer, designed to describe and explain the human and physical patterns on the earth as well as the interaction between people and the environment, systematically yield a unique perspective that is often ignored by political scientists, historians, and other analysts.

This book, then, is a geographer's guide to Iraq. Our goal is to offer a coherent but concise source of information about the country and its people. It will also serve as a point of departure for a more detailed study of the physical and cultural geographic features summarized here. Much has been written about Iraq, particularly since the 1991 Gulf War. What seemed to be missing was a synthesis of the data into a format that facilitated conceptual understanding of the critical issues that affect international security. This book is not designed to be an encyclopedic look at all aspects of Iraqi culture or Iraq's environment. It does, however, provide a systematic yet concise view of the physical and human environment of Iraq using methods of analysis common to our training as geographers.

METHOD OF ANALYSIS

Geographers have an assortment of tools at their disposal to analyze and characterize the earth. Our analysis of Iraq will be accomplished by applying a regional approach, the geographer's most important overarching method (Palka, 2001). Regions are spatial constructs that result from locating, grouping, and mapping places that share some common characteristic(s). For example, a region may be defined as the area inhabited primarily by a group of people speaking the same language, practicing the same religion, dressing in similar ways, or having the same economic base. The physical characteristics of a place may be equally important in helping to describe a region, such as grouping together areas with similar vegetation types. Thus, regions are defined based on some combination of physical and/or human features within an areal extent.

As indicated in Figure 1.1 [C-1], the regional analysis presented here will cover Iraq's location, geomorphology, climatology, biogeography, cultural and historical geography, economic geography, population and urban characteristics, and medical geography. In doing so, the geographers contributing to this book often draw upon methods and scholarship from other disciplines. Therefore, a cultural geographer might draw from anthropology or cultural studies to understand Iraqi language or religion, but ultimately that information is used to better explain the cultural patterns observed in Iraq. Although, as Figure 1.1 shows, a wide range of scholarship outside of geography will be used, ultimately the focus returns to Iraq's physical and human patterns. To better explain this perspective, it is appropriate to devote a few paragraphs here to explain the relevance of each chapter to the overall scope of the book. This will help to answer the "why?" or "so what?" of the material presented.

Any discussion of a country's geography must begin with a description of place, both in absolute terms as well as its location relative to other places. Chapter 2 therefore places Iraq in context of its position on the earth, within the region, and in relation to its neighbors. Although it often appears small on maps, Iraq is about the size of California and borders six important countries, including Iran, Saudi Arabia, and Turkey.

Geomorphology, climatology, and biogeography, covered in chapters 3, 4, and 5, are subfields of geography that help explain the environments upon which Iraqis interact on a daily basis. Once again, as indicated in Figure 1.1, these subfields of geography merge the scholarship of other disciplines, such as geology, meteorology, and biology, with a spatial perspective to better describe and explain physical patterns on the earth's surface. In Iraq, these patterns include large expanses of desert covering 40 percent of the country's land area, marshes, and towering mountains in the north. Water, when present, is vital, and it is easy to argue that without the life-giving Tigris and Euphrates Rivers, Iraq's history would be quite different. Human civilization owes much to these rivers and the rich soils they created on their banks.

In terms of weather and climate (chapter 4), Iraq is mostly dry and hot, although mountains in the north provide cooler temperatures and more rainfall. Because of the climate and terrain, natural hazards such as drought, flooding, sandstorms and dust storms, and earthquakes are common. Vegetation in Iraq (chapter 5) is sparse with notable exceptions in the northern mountains, in a narrow corridor along the Tigris and Euphrates Rivers, and in southern marshy areas near Basra.

In chapter 6, the focus shifts from the physical environment to the human landscape. Geographers emphasize that human culture modifies and is modified by the physical environment, and that is certainly true in Iraq. Historically, the waters of Mesopotamia yielded agricultural surpluses that allowed people to break from farming to pursue political, theological, scientific, and artistic goals. But the environment works against this region as well, and Iraq's location on a flat and easily crossed river valley subjected it to repeated invasions that brought new leaders and new cultural ideas. Arab, Persian, Turkic, and Hellenistic ideas are only a few of the cultural legacies of these invasions.

Iraq's diverse cultural landscape is addressed in chapter 7. Language and religion patterns make it clear that while many Iraqis share culture traits, there is also a great deal of diversity. Islam flourishes, but Sunni and Shiite divisions defy easy categorization. Kurd, Christian, Yazidi, and other minority groups have held on to their identity despite decades of periodic oppression.

Understanding Iraq's political history and geography is vital in a post-Saddam Iraq. Although Hussein's Baath Party dominated the country for over three decades, political opposition among geographically-based minority groups survived despite brutal government repression. In a freer Iraq, each of these groups will expect a say in the country's future. Chapter 8 addresses these political complexities.

Iraq relies on oil for its economic well-being and controls what some estimate to be the second-largest reserve in the world. As discussed in chapter 9, in spite of its oil and natural gas resources, Iraq's economy is weak, and even once-fertile agricultural areas have been hurt by water diversion schemes, salinization of soils, and rural-to-urban migration. UN sanctions added to the economic malaise. Now, Iraq must confront these problems to rebuild and move forward.

Additionally, Iraq's population continues to grow. At current rates, the country is expected to double in size in less than three decades. Today, more than 45 percent of Iraq's population is under age 15, a demographic time bomb that could present serious problems in the future. Other demographic statistics indicate that Iraq's population is worse off than its neighbors in terms of life expectancy, nutrition, and infant mortality. These issues are discussed in chapter 10. Iraq's urban centers are also outlined in this chapter, with a focus on the largest cities of Baghdad, Mosul, and Basra.

Finally, chapter 11 discusses health concerns in present-day Iraq, challenges sure to be of vital importance to any post-Saddam government. A variety of diseases and hazards that pose a risk include malaria, typhoid, tuberculosis, and cholera. Economic sanctions and a despotic government created a decline in the health of the average Iraqi during the 1990s.

CONCLUSION

Overall, we hope that it will become clear that a geographic perspective emphasizing generalized patterns on the landscape can provide a useful and practical framework for developing a better understanding of Iraq. As we complete this chapter of the book, we recognize that the situation in Iraq is changing dramatically in ways that make it difficult to predict the long-term implications to the physical and cultural landscape, the Iraqi people, and their way of life. Nevertheless, a geographic understanding of Iraq and its peoples is an absolute imperative for politicians, academics, military personnel, and nongovernmental agencies seeking to restore the peace and rebuild the country in the aftermath of war.

2

Location

Wiley C. Thompson

Key Points

- To understand Iraq's geography, it is first necessary to appreciate its absolute and relative location
- Iraq shares borders with six countries
- Iraq has limited access to the sea
- Iraq is slightly larger in area than California

One of the keys to understanding the importance and complexity of Iraq and its role in world affairs lies in its location. As geographers seek to answer the question of "where?" in their regional analysis, they examine the concept of location in two ways, that is, by considering absolute and relative location (Figure 2.1). Absolute location refers to the exact position of a place on the surface of the earth, and it is usually expressed in terms of latitude and longitude or some other coordinate system. For example, the geographic center of Iraq lies at approximately 33°N latitude, about the same as Tennessee, and 44°E longitude.

The north-south extent of Iraq's borders run from 37°21′N in the Kurdistan region along its northern border with Turkey to 29°04′N on its southern border with Saudi Arabia and Kuwait. This translates to roughly 900 km (560 mi) along the north-south axis. Iraq's east-west extent is about the same distance, running from 38°56′E in the Syrian Desert to 48°36′E in the vicinity of the Shatt al Arab on the Persian (or Arabian) Gulf (Figure 2.2 [C-2]). Two significant lines of latitude after the 1991 Gulf War were the UN-imposed 36th parallel "no-fly zone" in the north, intended to protect ethnic Kurds, and the 33rd parallel no-fly zone in southern Iraq, designed to protect the oft-persecuted Shia Muslims in the south.

Iraq is a fairly compact country in shape, with its geographic center relatively equidistant from its borders. Geographers generally regard this shape as being beneficial for a country because it makes administrative control, transportation, and economic integration relatively easy compared to countries that

Figure 2.1 The Location of Iraq

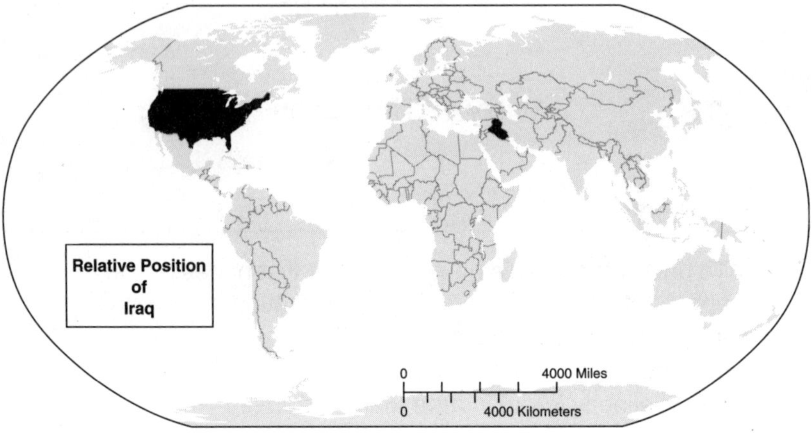

Relative Position
of
Iraq

0   4000 Miles

0   4000 Kilometers

Source: Wiley C. Thompson

are elongated or have irregular shapes. Notice that Baghdad lies quite close to the center of the country.

The second method by which a geographer answers the question, "where?" is with the concept of relative location. Relative location describes the location of a place in relation to the position of other places or things. The relative location perspective is affected by distance and accessibility to other resources or influences within a larger region or realm. A city may be physically close to another location but relatively far away if they are separated by a high mountain chain or a body of water. The relative location of a region is a key element in the analysis of the historical, cultural, political, and economic geography of a region. As such, only a cursory introduction to Iraq's relative location will be given now, saving more critical analyses for later chapters.

Iraq is a country of Southwest Asia, bordered by Syria and Jordan to the west, Turkey to the north, Iran to the east, and by Saudi Arabia and Kuwait to the south (Figure 2.3 [C-3]). This fact of political geography remains a crucial aspect of Iraq's relative location. Neighbors Turkey and Syria control parts of Iraq's important water sources, the famed Tigris and Euphrates rivers. Bordering countries also share ethnic groups with Iraq, a reality that has often led to tensions among the countries in the region. For example, Kurdish populations are divided primarily among Iraq, Turkey, and Iran. In addition, Shia Muslim populations straddle the border between Iran and Iraq.

Economic relationships can also be examined in terms of relative location. For example, Iraq's only significant access to the sea, and thus maritime shipping of oil and other products, comes from the Shatt al Arab, a short, 170 km (110 mi) river that flows from the confluence of the Tigris and Euphrates River to the Persian Gulf. The Shatt al Arab also serves as the border between

Figure 2.4 Size Comparison of Iraq and the United States

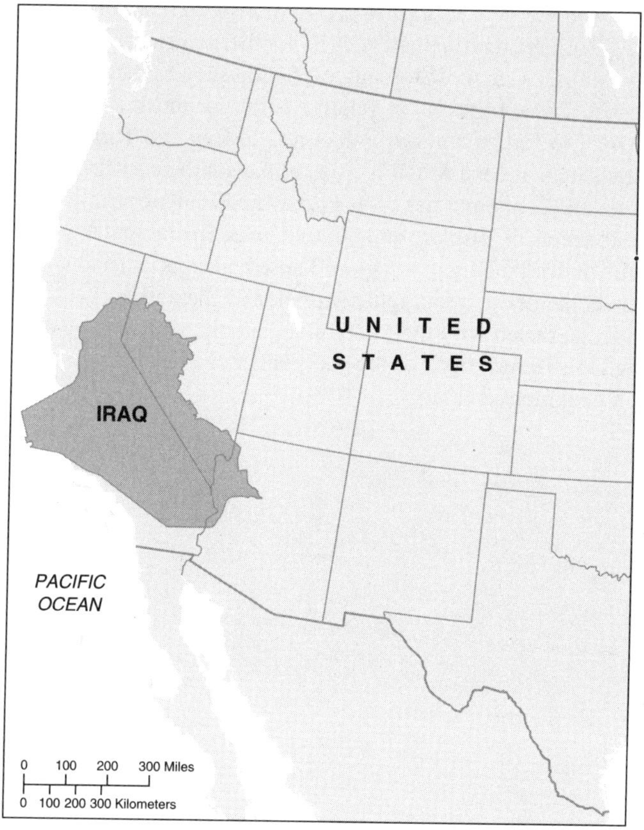

Source: Wiley C. Thompson

Iraq and Iran in this area and was the location of fierce fighting during the Iran-Iraq war of the 1980s as well as during the United States and United Kingdom–led conflict in 2003. Limited access to the sea also increases Iraq's dependence on trade through neighboring states such as Jordan and Syria.

With a total area of 437,065 sq km, Iraq is just larger than the state of California (Figure 2.4). The country has high, rugged mountains along its border with Iran and Turkey. Haji Ibrahim, along the Iran-Iraq border is the highest point in Iraq at 3,607 m (11,834 ft.). Iraq's elevation tapers to the south and west as it transitions through the fertile Tigris and Euphrates basins to the Syrian Desert and the red sand deserts of the An Nafud in northern Saudi Arabia. The rugged mountains of southern Turkey and northern Iraq are home to the Kurdish people. The mountains on Iraq's eastern border with Iran are home to Kurds and Shia Muslims. The satellite image in Figure 2.5 [C-4] provides a sense of the dramatic transition from desert plains to rugged mountains along the Iran-Iraq border.

In terms of relative location, the mountainous northern and eastern borders have conspired with history to create more differences between Iraq and neighboring Turkey and Iran than with its southern or eastern neighbors. Iraqis are largely Arabs, as are most Jordanians, Saudis, and Syrians, but Iranians and Turks are not. Thus, in terms of relative location, much of Iraq is, in many ways, "closer" to Jordan and Syria than it is to Iran and Turkey. But in some areas of Iraq, such as Shia Muslim cities in the south and Kurdish areas of the north, ethnic and economic ties with Turkey and Iran are strong.

The concepts of absolute and relative location are vital to a geographer's analysis and understanding of a region. These concepts can and will be applied to the various themes or geographic subfields as the authors focus on how the people have interacted with their neighbors and the surrounding environment. It is through this framework that the geographer can better explain the physical and human geography of Iraq.

3

Geomorphology

Matthew R. Sampson

Key Points

- Iraq has four major physiographic regions: desert, uplands, highlands, and an alluvial plain
- Deserts cover over 40 percent of Iraq's area, especially in the western and southern parts of the country
- The Tigris and Euphrates Rivers, the lifeblood of Iraq, are vital to human populations and agricultural activity

Geomorphology is the study of landforms and the processes that shape them. It entails understanding the terrain, or the lay of the land, and how it got that way. The geomorphic processes that shape the earth's surface include weathering and erosion. The former breaks down surface materials either by physical or chemical means. The latter refers to the movement of weathered surface material by wave action, running water, wind, or glacial ice. The result is a dynamic landscape that is continually being reshaped by the forces of nature. Depending on the dominant geomorphic forces at work, distinct physical landscapes will result. If running water is the dominant force, for example, it would create specific landforms such as river valleys, alluvial fans, and deltas. Geographers often group these distinct landscapes into geographic regions.

GEOGRAPHIC REGIONS

While the territory of Iraq consists primarily of lowlands that seldom exceed 300 m (1,000 ft.) in elevation ("Iraq Land & Climate," 2002), the country can be divided into four major geographic regions (Figure 3.1 [C-5]): desert in the west and southwest; rolling uplands between the Euphrates and Tigris rivers; highlands in the north and northeast; and an alluvial plain in the central and southern areas (Library of Congress, 1988).

Deserts

West of the Euphrates lies a vast extension of the Syrian Desert that covers parts of Syria, Jordan, and Saudi Arabia beyond Iraq's borders. In the desert, stony plains give way to islands of sand. Elevations can reach to over 450 m (1,500 ft.) in some areas. Western deserts are furrowed with numerous *wadis*, riverbeds that are normally dry and that direct occasional rainfall east toward the Euphrates (Held, 2000). Wadis, which can stretch to over 400 km (250 mi) in length, can suddenly become raging rivers during periods of rainfall.

The deserts of southern Iraq are divided into the al-Hijarah in the western area and al-Dibdibah in eastern sections. The al-Hijarah Desert is a rough, uneven limestone plateau cut by wadis and depressions and strewn with loose flint and chert, a flint-like quartz. The al-Dibdibah is more sandy and gravelly than the al-Hijarah. A dominant feature of the al-Dibdibah is the Wadi al-Batin, which runs along the Iraqi border with Kuwait. West of the Wadi al-Batin is a flat open stretch of desert that was utilized by Coalition forces in February 1991 to flank Iraqi troops during the first Gulf War (Held, 2000).

Uplands/Al Jazirah

North of the Tigris city of Samarra and the Euphrates city of Hit is an uplands region known as al-Jazirah, "the island" (Library of Congress, 1988). This area is a desert plateau that extends into Syria. The Wadi Tharthar is the main source of drainage for the region, running into the large Tharthar Depression, which has been improved over the years to include a large artificial lake that can accept floodwaters from the Tigris if necessary (Held, 2000).

Highlands

In northern Iraq, highlands give way to mountains that rise to over 3,600 m (11,000 ft.) near the Turkish and Iranian borders. Iraq's highest point, at 3,607 m (11,834 ft.) is Haji Ibrahim, located on the Iranian border.

In general, the highland ridges in the north extend east–west, which follows folds in the Taurus and Anti-Taurus mountains in neighboring Turkey. In the northeast these folds turn toward the southeast as part of the mighty Zagros Mountains that primarily lie across the border in Iran. These highlands are natural headwaters for streams such as the Khabur, Great and Little Zab, Udhaym, and Diyala that flow west to join the Tigris in its journey south. Much of these highland areas are still remote and relatively inaccessible. Few passes cut through the Zagros, but notable gaps include the Spilak Pass in the Ruwanduz River gorge, the Ali Beg Gorge west of the Ruwanduz, and the Shinak Pass (Held, 2000).

Alluvial Fan

A huge alluvial plain begins north of Baghdad and extends to the Persian Gulf. Alluvium, and thus the term *alluvial*, simply refers to sedimentary material, such as sand and silt, deposited by flowing water. Over the centuries, silt carried in the Tigris, Euphrates, and other rivers was deposited in their deltas.

With time, the delta areas grew southward, and today much of southern Iraq is composed of the sand from eons of deposition.

This area is generally low and flat, ranging from about 25 m (80 ft.) to below sea level in some areas (Held, 2000, 337). Channels of the Tigris and Euphrates as well as irrigation canals cut across the area. During times of flooding, lakes and large muddy areas can also be found in this region. Near the confluence of the Tigris and Euphrates at al Qurnah is a marshland known as the Hawr al Hammar, which historically extended eastward into Iran (Library of Congress, 1988).

In 1992 the Iraqi government engineered a 563 km (350 mi) artificial river known as the Main Outfall Drain (MOD), nominally to improve drainage in this area (see Figure 3.2 [C-6]). "Sometimes termed the Third River, the MOD is designed to drain both the central marshes and the southern marsh homeland of the Mandan" (Held, 2000). The Mandan, "Marsh Arabs," mostly Shiite Muslims, were seen as a threat by the Hussein government. As the marshes dried up, thousands of Marsh Arabs fled to ethnically similar areas in Iran. Figure 3.3 [C-7] illustrates marsh decline over the past three decades.

High salinity levels in the soil are also an issue in this area. In addition to salts naturally found in the Tigris and Euphrates, irrigation schemes and poor agricultural practices can cause elevated salt levels in the soil. A high water table and poor surface drainage complicate the problem. In general, soil salinity increases south of Baghdad to the Persian Gulf and severely limits agriculture in the region south of al Amarah. A large, usually dry lake southwest of Baghdad, known as the Bahr al Milh, or Sea of Salt, is indicative of the salinity in the alluvial plain (Library of Congress, 1988).

Rivers

Rising in Turkey, the Euphrates is joined by the Nahr River in Syria before crossing into Iraq. The Tigris also begins in Turkey, but is significantly augmented by several rivers above and one river below Baghdad. The two rivers meet at al Qurnah. Below their confluence, both rivers break into many smaller channels and drain into marshlands in the area. As mentioned earlier, large amounts of silt are deposited here. The rivers drain into the Shatt al Arab, which is repeatedly filling with silt from the Karun River that enters from Iran. This silting problem affects the navigation of ocean-going ships attempting to reach the port of Basra (Library of Congress, 1988).

Although these mighty and historic rivers bring life to Iraqis, over the centuries they have also brought death and hardship in the form of flooding. Before the 1980s, March, April, and May often saw a forty-fold increase in river volumes, but dams and diversion projects in Turkey have now greatly reduced the amount of water entering Iraq. September and October are the months when the rivers are at their lowest.

Geology

Geologically, the Middle East is a particularly complex region (Beaumont, Blake, & Wagstaff, 1988). In terms of plate tectonics, northeastern Iraq and western Iran is a convergent boundary where plates of the earth's crust come together. Specifically, the region lies along the impact zone of the Arabian and Eurasian plates (*Goode's World Atlas,* 2001). Both plates consist of continental crust, and the northward movement of Arabia toward Eurasia has produced widespread overthrusting and folding (Beaumont, Blake, & Wagstaff, 1988).

> The result is a continuous band of folded and faulted mountains from western and eastern Anatolia and then southeastward across Iran and eastward into the Himalayas. This belt makes Turkey, northeastern Iraq, and Iran structurally extraordinarily complex. (Held, 2000)

This convergent zone also causes earthquakes, which in Iraq are almost entirely in the northeast region. As if to counter the curse of regular earthquakes, the region's geology has also blessed Iraq with petroleum resources. Almost all early finds of oil were made in the sedimentary basins where the Zagros Mountains meet the plains (Beaumont, Blake, & Wagstaff, 1988). Oil will be discussed in greater length in the chapter on Economic Geography later in this volume.

Soils

As expected, desert soils are common in Iraq. These light gray or brown soils have low levels of organic matter and overlie limestone and other calcium-containing rock. Although these soils normally support only shrubby desert plants, cultivation can be successful when irrigation is present (Held, 2000).

In areas watered by the Tigris and Euphrates fertile, alluvial soils can be found. But, as mentioned above, salinity from excessive irrigation and inadequate drainage have damaged the fertility of the region. Most plants are salt-intolerant and will not grow if the salt concentration in the soil gets too high. Consequently, salinization can have a devastating effect on agricultural production.

Summary

Iraq is mainly a low-lying country with a small area of high relief in the northeast. The western third of Iraq consists of desert terrain dissected by wadis, but the principal geomorphic feature of Iraq is the alluvial plain formed by the Tigris and Euphrates Rivers and stretching from north of Baghdad to the Gulf. These rivers and their sediment have shaped the land and life of Iraq since ancient times, and they continue to do so today. Arguably, the Tigris and Euphrates are Iraq's most important resource, providing precious water and fertile soil. In an arid region such as Iraq, water is life; consequently, the Tigris and Euphrates are the lifeblood of the country.

4

Climate

Richard P. Pannell

Key Points

- Iraq has three distinct climatic zones: subtropical desert, subtropical steppe, and dry summer subtropical
- Iraq's deserts are extremely hot and dry in the summer and cool and dry in the winter
- Precipitation is limited, with even the wettest areas getting less than 30 inches per year
- Cold is a significant hazard in the northern mountains
- Sandstorms and dust storms are common, especially in the deserts
- Spring flooding is a real hazard in many areas

The climate of Iraq must be characterized with some care. On the one hand it is simplistically predictable: summers are almost guaranteed to be very hot and extremely dry, while winters are typically characterized by mild to cool temperatures with precipitation. Yet the potential for a variety of climatic or weather hazards such as drought, sandstorms, and floods makes the reliance on short- and long-term forecasts problematic. This analysis will examine three major climate regions as well as the potential for natural disasters related to hydrologic or atmospheric phenomena.

Because of its location between 29° and 37°N latitude, Iraq is strongly influenced in the summer by subtropical high pressure. This high-pressure zone influences desert regions across North Africa and the Arabian Peninsula and migrates northward in the summer because of increased solar radiation associated with the summer solstice. High pressure is characterized by air masses that are descending toward the ground. Air moving downward is almost always dry because water vapor gets warmer toward the earth's surface and therefore does not condense to form rain.

By contrast, during the winter solstice, as the northern hemisphere is tilted away from the sun, the subtropical high pressure is replaced by periodic low-pressure systems that travel from west to east across Iraq bringing winter rains and snow in the mountain regions of the north. Low-pressure areas are characterized by rising air. As air rises, it cools, and water vapor condenses to a liquid state, yielding more precipitation.

15

These annual migrations of subtropical high pressure and midlatitude low pressure are clearly reflected in Iraq's climate, which can be classified into three distinct climate regions using the Köppen climate classification scheme (as simplified by Trewartha), a system commonly used by geographers. At its most basic, the Köppen system divides climates into Tropical/Rainy climates (designated with the first letter "A"), Dry ("B"), Humid and Mild ("C"), Humid and Cold ("D"), and Polar ("E") (Goode's, 2000). Within each lettered category, additional subtypes are identified, so a truly dry desert might be "BW" while a semidry steppe would be given the designation "BS." The "B" indicates a dry climate, while the "W" or "S" provides more detail about what type of dry climate. Sometimes, as seen below, a third letter can also be used to represent subtler differences.

The southern half of Iraq from the coastal areas near Basra to the Syrian Desert is closest to the subtropical high-pressure zone and consequently is classified as a Subtropical Desert ("BWh" in the Köppen system). The upland region north of Baghdad is significantly wetter, particularly in winter, and can be classified as a Subtropical Steppe ("BSh"). Finally, in the northern mountain regions, where conditions are much cooler and rainfall more abundant, the climate is Mediterranean or Dry Summer Subtropical ("Csa"). Mediterranean climates refer to areas such as, but not limited to, the Mediterranean Sea region that have hot, dry summers and cooler, wetter winters. The climate map in Figure 4.1 outlines these three regions, and each is shown in greater detail in the climographs (Figures 4.2–4.4).

SOUTHERN IRAQI DESERT

The arid landscape from the edges of the Syrian Desert in western Iraq to the coastal lowlands in the southeast is inhospitable to say the least. This Subtropical Desert climate ("BWh") is primarily controlled by the extended presence of high pressure throughout the year. This results in only small amounts of precipitation during the winter months when high pressure is occasionally displaced southward. Temperatures are extremely hot in the summer and mild in the winter due to variability of incoming solar radiation (insolation) caused by the tilt of Earth's axis. This annual variability can be clearly seen in the climographs of Baghdad, Basra, and Rutbah (Figure 4.2). Climographs show average precipitation and temperature by month. These graphs do not show the entire story, however, as mean daily maximum temperatures are typically 43°C (110°F) during the summer months and mean daily minimum temperatures are in the range of 5–10°C (40–50°F) during winter. In addition to the variability of insolation in Iraq, its location astride the African and Asian landmass results in a continental influence indicated by broad annual as well as daily (diurnal) temperature ranges. In other words, the region can experience wide temperature swings each day and over the course of a year. This "continentality" is primarily manifested through the rapid heating and cooling of the land surface due to its relatively low specific

Figure 4.1 Regional Köppen Climate Classification of Iraq

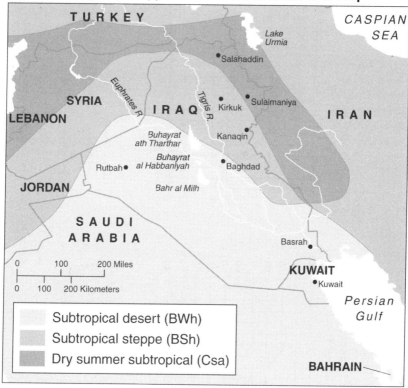

Source: Base map after Thompson, 2002; Climate data from Air Force Combat Climatology Center, 2001

heat (McKnight, 2002). In other words, the land heats up quickly and cools down fast when compared to the oceans. Countries along oceans often see less dramatic yearly and daily temperature swings because water tends to heat and cool slowly, providing a moderating effect on temperatures.

The other important climatic variable in this region is the wind (see section on hazards below). The dominant wind pattern for most of Iraq is the *shamal* (Arabic for "north"). Throughout the year, winds from the north and northwest bring very dry air across the country, which inhibits cloud development and thus precipitation. Average daily wind speeds are in excess of 10 knots during summer months, causing frequent dust storms and sandstorms, while winter conditions often generate faster but less frequent winds (AFCCC, 1997). In the early summer and early winter a different wind system, the *sharki* (Arabic for "east"), may occasionally blow from the south or southeast. June is typically the worst month for this type of activity due to the combined frequency of shamals and sharkis in central Iraq.

Despite the lack of moisture, much of this region is under cultivation, particularly the lowlands adjacent to the Tigris and Euphrates Rivers, where

Figure 4.2 Köppen Climographs for Baghdad, Basra, and Rutbah

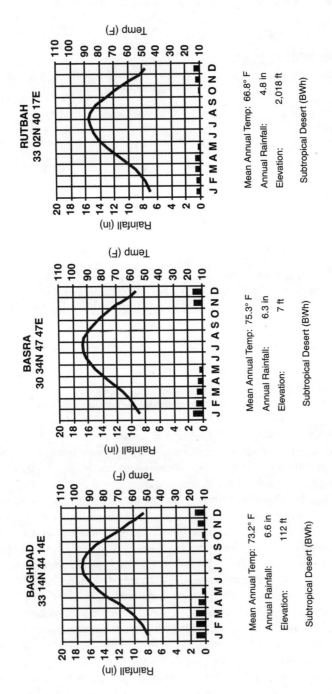

BAGHDAD
33 14N 44 14E

Mean Annual Temp: 73.2° F
Annual Rainfall: 6.6 in
Elevation: 112 ft

Subtropical Desert (BWh)

BASRA
30 34N 47 47E

Mean Annual Temp: 75.3° F
Annual Rainfall: 6.3 in
Elevation: 7 ft

Subtropical Desert (BWh)

RUTBAH
33 02N 40 17E

Mean Annual Temp: 66.8° F
Annual Rainfall: 4.8 in
Elevation: 2,018 ft

Subtropical Desert (BWh)

Source: Air Force Combat Climatology Center, OCDS, 1995

annual flooding has created rich alluvial soils. Although conditions are arid, irrigation canals and water diversion projects have allowed the people to farm this land for millennia. The potential for long-term drought is greatest here, however, and as recently as 1999–2000 this region experienced a hundred-year drought, meaning that on average a drought this severe only occurs once every century (NOAA, 2000).

IRAQI UPLANDS

North of the Tigris and Euphrates River valleys, the landscape rises gently toward the foothills of the Zagros Mountains. This region of Iraq is only slightly cooler but significantly wetter, resulting in its classification as a Subtropical Steppe ("BSh"). Summer temperatures are still extreme, but winter temperatures are slightly cooler. The continental influence is greater here, affecting both annual and diurnal temperature ranges. Average annual temperature ranges are over 25°C (50°F), while average daily temperature ranges are between 10° and 15°C (20–30°F) (AFCCC, 1997. Once again, migrating pressure patterns result in dry high pressure dominating during the summer months and wetter low-pressure systems occurring more frequently in the winter. Figure 4.3 shows climographs for Kanaqin and Kirkuk, two climate stations in this region.

Figure 4.3 Köppen Climographs for Kanaqin and Kirkuk

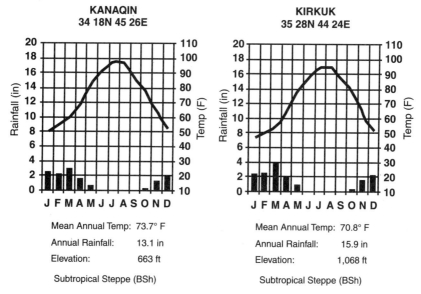

Mean Annual Temp: 73.7° F
Annual Rainfall: 13.1 in
Elevation: 663 ft
Subtropical Steppe (BSh)

Mean Annual Temp: 70.8° F
Annual Rainfall: 15.9 in
Elevation: 1,068 ft
Subtropical Steppe (BSh)

Data Source: Air Force Combat Climatology Center, OCDS, 1997

Increased moisture in this region promotes grassland and shrub-form vegetation and the development of organically rich soils suitable for agriculture. Likewise, the increased potential for precipitation makes flash flooding a potential hazard, particularly when thunderstorms can produce two to three inches of rain in less than 24 hours. The potential dangers of flash flooding are greatest in wadi (dry riverbed) bottoms where flows are quickly concentrated and soil conditions may be transformed into a type of quicksand for up to 24 hours (AFCCC, 1996).

THE NORTHERN MOUNTAINS

The final climate region of Iraq is the northern mountains, where conditions are much different than the rest of the country. Because of cooler temperatures and increased precipitation, this region is classified as a Mediterranean or Dry Summer Subtropical climate ("Csa"). Higher elevations in this area result in significantly cooler temperatures. Temperature differences between summer and winter are great here because it is far from the moderating effects of the ocean. In winter, temperatures are often below freezing and snow is common at altitudes above 1,000 m (3,300 ft.) with a thick snowpack in places above 1,500 m (5,000 ft.) (AFCCC, 1997). The annual rainfall for most of this region is in excess of 640 mm (25 in). In the winter precipitation is caused by migrating low-pressure systems from the Mediterranean Sea as they track across northern Iraq and rise over the mountains. The result is heavy rain or storm activity during the winter and spring, with precipitation being common from November through May. This pattern is shown in the climographs of Salahaddin and Sulaimaniya depicted in Figure 4.4.

The primary climate hazards in this region are related to the cold, wet conditions in winter. High mountain regions may be impassable due to snowpack, and roads are often icy and foggy. Snowstorms generated from cool, moist Mediterranean air may develop with little warning. Rapid, unexpected temperature changes make hypothermia and frost-bite a potential problem. Downslope winds called *föehns* are also a hazard and may strike without warning with wind gusts in excess of 50 knots (AFCCC, 1997).

CLIMATE AND WEATHER HAZARDS

Associated with the three general climate regions are several natural hazards that occur on a variety of timescales. Over longer periods of time, such as a 10-year timescale, conditions such as drought may be a significant anomaly to the normal climate regime. At the annual timescale, snowmelt and frontal activity influence large-scale flooding, while pressure patterns influence prevailing wind conditions. At the seasonal or even daily timescale, wind systems are generated that produce sandstorms and thunderstorm activity capable of causing rapid, violent flash flooding.

Figure 4.4 Köppen Climographs of Salahaddin and Sulaimaniya

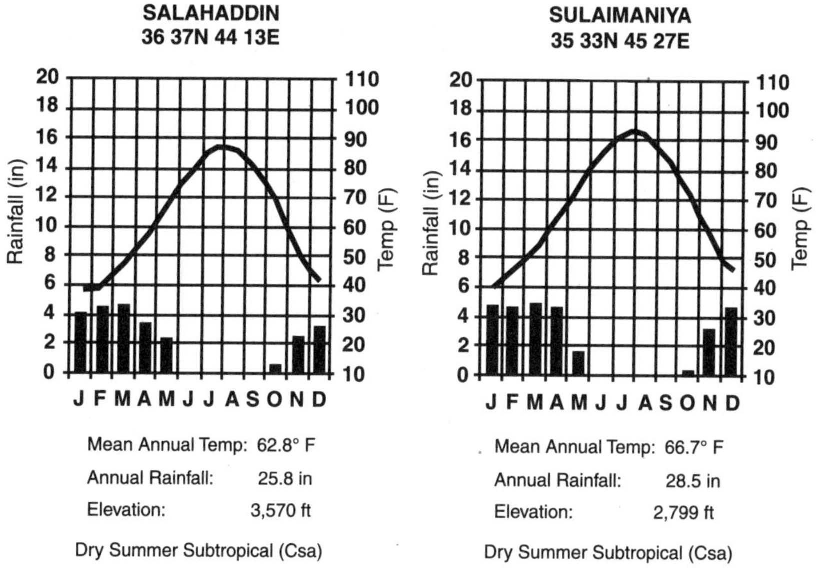
SALAHADDIN
36 37N 44 13E
Rainfall (in)
Temp (F)
J F M A M J J A S O N D
Mean Annual Temp: 62.8° F
Annual Rainfall: 25.8 in
Elevation: 3,570 ft
Dry Summer Subtropical (Csa)
SULAIMANIYA
35 33N 45 27E
Rainfall (in)
Temp (F)
J F M A M J J A S O N D
Mean Annual Temp: 66.7° F
Annual Rainfall: 28.5 in
Elevation: 2,799 ft
Dry Summer Subtropical (Csa)

Data Source: Air Force Climatology Center, OCDS, 1997

Drought

The importance of water for agriculture cannot be overemphasized in Iraq. Drought conditions affect not only the potential for rain-fed agriculture, but also alter the hydrologic conditions of the Tigris-Euphrates River system, a vital resource for irrigation. The occurrence of drought is related closely to two pressure systems: migrating low-pressure systems from Europe and the strength of Siberian high-pressure systems during the winter months. Droughts occur as the result of decreased precipitation in the winter and spring, which in turn is linked to the strength of Siberian high pressure. The stronger the Siberian high, the more likely westerly low-pressure systems will be blocked from reaching the interior of Iraq. In addition, warmer surface temperatures evaporate much of the available surface water, exacerbating conditions.

During the last 20 years, numerous droughts have occurred in Iraq, including 1984, 1990 and most recently, 1999–2000. While conditions have improved recently due to increased precipitation, the potential for drought remains a significant climatological concern for the people of Iraq, particularly in the arid regions of the southern half of the country.

Flooding

Flooding is a potential hazard on two different timescales. Every year, flooding of the Tigris-Euphrates River valley can be expected during the spring because of snowmelt and increased precipitation. The magnitude of these floods is related to the amount of snowpack in the northern highlands as well as the intensity and duration of spring precipitation in the northern half of the country. Timing of this event varies, but can be predicted to occur in March or April, typically the wettest months. River depths can be in excess of six m (20 ft.) across the floodplain, but velocity is relatively slow at about two knots (AFCCC, 1996).

At much shorter timescales, the potential for flash flooding is a separate hazard that occurs primarily in winter and spring, but is more unpredictable. Thunderstorms though uncommon can result in brief, intense rainfall events. Because surface conditions are largely unvegetated, runoff quickly moves through the drainage system creating high discharges in wadis. These flash floods may occur almost anywhere, but typically last only a few hours.

Sandstorms

The presence of poor, dry soils combined with continuous winds during most of the year make the potential for dust storms and sandstorms an important hazard to consider. This is especially prevalent in the southern deserts where vegetative cover is at a minimum and the supply of silt and sand is relatively abundant. Even in more humid regions adjacent to the Tigris and Euphrates Rivers, however, this is becoming a problem due to poor irrigation practices, which cause desertification.

As previously discussed, most of Iraq is dominated by the shamal. This steady wind blows throughout the country from about 8 to 10 knots on average, but is capable of much faster gusts, when wind speeds can exceed 50 knots locally. The shamal is strongest in the winter and the Air Force Combat Climatology Center (AFCCC) refers to two different winter shamals: the 24–36 hour shamal and the 3–5 day shamal. The 24–36 hour shamal is associated with the passage of a front and is characterized by fast winds (25–30 knots) immediately behind the front, blowing from the north or northwest. These winds, capable of generating dust storms and sandstorms, are relatively common, occurring two or three times a month during the winter season (AFCCC, 1996). The scene is captured in Figure 4.5 [C-8]. The 3–5 day shamal, on the other hand, is not associated with frontal passage, but, rather, it is caused by the stalling of pressure systems over the Strait of Hormuz, which causes an increased pressure gradient between high pressure from the north and low pressure over the Persian Gulf. These systems are much more intense but occur less frequently, arising only a couple of times during the entire winter season. It is from these shamals that the strongest sandstorms are generated with sustained winds of 25 knots for several days (AFCCC, 1996).

During the transition months in early summer and early winter, another type of wind, the sharki, blows from the south or southeast. This wind is also associated with the passage of low-pressure systems; however, it is derived from warm, dry air moving ahead of the front. These winds originate from a continental tropical air mass over the Arabian Peninsula and are similar to the Santa Ana winds of the southwest United States. In Iraq they promote the desiccation of land and wilting of plant life (Held, 2000). This wind is particularly common in June, when continental air is gaining strength but migrating low-pressure systems are still possible.

SUMMARY

Iraq's climate displays clear latitudinal zonation that can be related directly to migratory surface pressure zones. The southern half of Iraq is essentially a vast subtropical desert region across its entire east-west expanse. Conditions become moister from south to north as the potential for migrating low-pressure cyclonic activity increases. This results in a subtropical steppe environment in the uplands north of the capital city, Baghdad. Finally, in the higher elevations in the northernmost reaches of the country a longer wet season occurs due to increased frontal activity and orographic uplift. Cooler temperatures prevail due to altitude and the region has a mild, humid Mediterranean climate with a dry summer season.

Despite these distinct climate zones a variety of environmental hazards associated with anomalous climate and weather patterns result in droughts, floods, and sandstorms. The long-term variability in moisture conditions can result in periods of extended drought, like the one Iraq faced during 1999–2000. On an annual timescale, floods associated with spring rains and snowmelt in the northern highlands results in large-scale flooding of the Tigris and Euphrates River basins of central Iraq. While on a shorter timescale thunderstorm activity can create flash floods in wadis. Finally, periodic and seasonal wind patterns create two wind patterns, the shamal and the sharki, both capable of generating dust storms and sandstorms lasting hours to days.

5

Vegetation and Soils

Peter G. Anderson

Key Points

- Iraq is largely devoid of significant natural vegetation
- Minor forest resources, already significantly impacted by human activity, exist in the northern mountains
- The most unique natural biome in Iraq is the marshland area of the lower Tigris and Euphrates, but this area was damaged by Iraqi government policies in the 1990s that aimed at reducing populations in the region

The natural vegetation of Iraq falls into two categories: 1) sparse to none; and, 2) altered by human activities. Roughly 74 percent of the land area exhibits a medium to high degree of human disturbance (Animal Info, 1999). The "gift of life" brought to Iraq via the Tigris and Euphrates Rivers influences an extensive area. The water of these rivers has been harnessed by the area's inhabitants for millennia, and the natural ecosystems of the river landscape have been altered repeatedly. Because rainfall throughout Iraq is minimal and deserts are common in western, southwestern, and southern Iraq, little grows in these regions. Although rainfall in the mountains of the north is sufficient for forest development, the spatial extent of this area in Iraq is small. Iraq may be blessed with warmth and petroleum, but the country is depauperate in natural vegetation and ecological diversity. About one-third of the country is desert, one-third grass and shrub, and one-third cropland and settlements (Animal Info, 1999).

SOILS

Much of Iraq is arid to semi-arid, receiving less than 30 cm (12 in) of rainfall per year, and centuries of human use have altered Iraq's ground cover and soils. These factors contribute to its poor soil quality, affecting agricultural potential. The only soils with a satisfactory potential for agricultural production occur in the Tigris and Euphrates lowlands. Soils of this region may have abundant clay,

25

humus, and ground water, creating a rich soil for agriculture (Animal Info, 1999). Flood control and irrigation canals direct water to where human interests want and will use the water. Unfortunately, decades and centuries of topsoil loss has degraded Iraq's good soils, while dams on the rivers have altered spring floods that once replenished floodplain environments.

VEGETATION REGIONS

Iraq presents three main vegetation types, which are discussed below: mountains in the north, desert in the south, and central lowlands, the valley between the Euphrates and Tigris Rivers.

Northern Mountains

Northern Iraq consists of hilly to mountainous terrain. Elevations rise to greater than 1,800 m (6,000 ft.) near the Turkish border. Along the Iranian border, peaks greater than 3,050 m (10,000 ft.) may be found, such as Mount Halgurd and Haji Ibrahim. The spatial extent of this region in Iraq is relatively small, confined to areas near the borders of Turkey and Iran. This highland area experiences the four seasons, with cool dry summers and cold snowy winters. Rainfall in the northern highlands occurs from October to May, ranging from 30 to 56 cm (12–22 in) (Microsoft, 2002). Spring melt of the winter snows feeds the Tigris and Euphrates Rivers and their tributaries, such as the Great Zab, Little Zab, and the Diyala.

Eight to ten thousand years ago, the highlands region was heavily forested with cedar, pine, juniper, ash, poplar, sycamore, chestnut, and oak (Izady, 1997). Due to climate change and centuries of human use, oak is the only predominant tree of the region today ("Kurdish Land," 2002). The loss of a protective vegetative cover has contributed to increased soil erosion and local climate change, and the moist microclimate of shaded woodland is now a dry landscape. Although all of the vegetation in the highlands region of Iraq has experienced some degree of human use and alteration, the pasture lands remain in reasonably good condition, and nomadic herding and subsistence agriculture is a prevalent economic livelihood in the region ("Kurdish Land," 2002).

Southern Deserts

Whereas Iraq's northern mountainous region is a small area, lowlands (<610 m or 2,000 ft.) and aridity dominate most of the rest of the country. Lands less than 300 m (under 1,000 ft.) can be found in the basin of the Tigris and Euphrates Rivers. These rivers flow northwest to southeast. Lands north and south of the river valleys are hot, dry deserts. Iraq's southern and western lands, contiguous with Saudi Arabia, Jordan, and part of Syria, are an extension of the Syrian Desert. The environment here may be best described as bleak. Broad plains of sandstorms and dust storms are the result of long hot, dry summers and short cool winters (Microsoft, 2002). During summer, daytime temperatures in the Syrian Desert may exceed 49°C (120°F), while nights may cool to less than 10°C (50°F). Little or no precipitation falls. This combination creates an environment where plant growth and vegetation development is minimal to nonexistent. Without irrigation, the desert is not arable and only sparse cattle and camels are raised by nomadic herders (Microsoft, 2002).

Central Lowlands

Iraq's central lowlands, the ancient land of Mesopotamia, formed the eastern end of the Fertile Crescent. In warm weather with good soils and copious groundwater, abundant plant growth occurred there. The lowlands persist, but the rivers have been harnessed for flood control, hydroelectric power, and irrigation purposes. As groundwater has been altered, soils have degraded and agriculture has replaced the natural vegetation and ecosystems. Although these activities are contributing to increased agricultural productivity, this once-verdant natural area is diminishing in size and quality (Animal Info, 1999).

The central lowlands, the Tigris and Euphrates River valleys and floodplains, consist of an upper and lower section. The upper section is centered around Baghdad, whereas the lower section encompasses lands of the southeast, once an extensive marshland region. Unfortunately, as with marshes and swamps the world over, this biologically productive habitat rapidly diminished at the hands of Saddam Hussein's government. As the wetlands disappear, so too does the plant and animal life of the region (see Figure 3.3 [C-7]).

The once-thriving wetland of this area included freshwater swamps and marshes of the Tigris and Euphrates Rivers, the Shatt al Arab, and the delta coast marshes of the northern Persian Gulf. "This complex of shallow freshwater lakes, swamps, marshes, and seasonally inundated plains between the Tigris and Euphrates Rivers is among the most important wintering areas for migratory birds in Eurasia. Of the 278 species of birds that have been recorded in the lower Mesopotamia, 134 species are dependent to some extent on the wetland habitats. These marshes support almost the entire world population of two bird species, the Basra reed warbler (*Acrocephalus griseldis*) and Iraq babbler (*Turdoides altirostris*)" (WWF "Global 200," 2002).

Unfortunately, these wetland areas are rapidly diminishing and may have been degraded to such a degree, by stream channelization, irrigation withdrawals, chemical contamination, agricultural, industrial, and residential pollution, and drainage for conversion to other land uses, that they have ceased to function as a natural ecosystem (Izady, 1997; WWF "Wetlands," 2002). Salinization is also a problem because of increased salt accumulation caused by the flushing of salts from agricultural lands (WWF "Wetlands," 2002; USGS, 2001).

In the wetlands of lower Mesopotamia, recent military activity has contributed to the loss of plants and animals, as well as human inhabitants (WWF

"Wetlands," 2002). "Much of the fighting during the prolonged Iran-Iraq War (1980–1988) and bombings during the Persian Gulf War occurred in and around the wetlands and caused considerable damage to the marsh ecosystems" (WWF "Global 200," 2002). Additionally, the marshes may have been purposefully drained as part of the "attempted elimination" of political opposition to Saddam Hussein's government (USGS, 2001). The net effect of these activities is the loss of one of southwest Asia's most productive wetland ecosystems.

CONCLUSION

Approximately one-half of Iraq may be considered arid to semi-arid lands with little to no vegetation and sparse animal life. The other half of Iraq's lands has been highly altered due to centuries of human activity and a recent assault on Iraq's landscape. The result is that scarce or no natural ecosystems remain within the geopolitical boundary referred to as Iraq, a country once rich with human and natural heritage that might now be described as ecologically bankrupt.

6

Historical Geography

James B. Dalton

Key Points:

- Iraq's history dates back to one of the earliest civilizations, in Mesopotamia
- Iraq's history is marked by periods of prosperity interspersed with eras of conquest and destruction
- Saddam Hussein came to power in the 1970s with the support of Sunni Muslims, who had been relatively favored since the times of British control

This chapter will highlight some of the key moments in the history of the area that is modern Iraq. The primary focus will be on the societies and a few of the leaders/rulers that made significant impacts. The chapter will discuss ancient Mesopotamia, the Muslim Arab period, the Ottoman era, the British period, independent Iraq, and the Saddam Hussein period. It is important to note that even though the flow from one historical period to another may be presented as if there were abrupt and immediate changes, this is not the case for the most part. Except during times of invasion, the transition between eras was often quite gradual.

ANCIENT MESOPOTAMIA

The area that today is Iraq and Iran has been inhabited for at least 10,000 years, and in many ways the history of early Iraq is the history of early human civilization. The first inhabitants were "vigorous, progressive, and aggressive" in all aspects of their lives (Polk, 1991). The remains they left behind tell of a people who irrigated the land, cultivated plants, and built walls around their villages for protection. In fact, by the fourth millennium B.C., cities had appeared on the plains of the Tigris and Euphrates valleys. Irrigation and the widespread domestication of plants and animals created an agricultural abundance, and these surpluses allowed people to pursue activities other than farming, such as politics, religion, and the arts. Excess agriculture also spawned trade, which re-

29

Figure 6.1 Cuneiform Writing

Source: Layard, A.H. 1849. *Nineveh and Its Remains.* New York: George P. Putman

quired merchants, transportation providers, security forces, and public officials such as tax collectors. Communities also needed good political organization to maintain common resources, such as irrigation ditches, and to ensure that the agricultural sector remained strong.

The third millennium B.C. was dominated by the Sumerians, with power focused in important cities such as Ur and Kish. The Sumerians are remembered for developing the form of writing known as *cuneiform*, a complex series of markings in clay tablets. Cuneiform, a form of which is depicted in Figure 6.1, was used throughout the Middle East for roughly two thousand years. At its height, the Sumerian Empire extended from the Zagros Mountains west to the Mediterranean Sea and from the Taurus Mountains in the north to the Persian Gulf, but was concentrated in the southern Tigris and Euphrates valleys, as indicated in Figure 6.2.

With time, Sumerian power waxed and waned, and by 2300 B.C., the region was taken over by Sargon I, a Semitic ruler. Sargon's people intermarried with the Sumerians to create a group known as the Akkadians. Consequently, this period is sometimes labeled Sumer and Akkad. The Akkadians extended the reach of their empire to include the entire Tigris and Euphrates watershed (Figure 6.2). But by about 2200 B.C. the region had been overrun by the Gutians, a group from the Zagros Mountains to the east. Within a hundred years Gutian rule had been overthrown and a period of social and cultural prosperity followed that saw important developments in legal codes and systems of education.

Figure 6.2 The Sumerian and Akkadian Empires

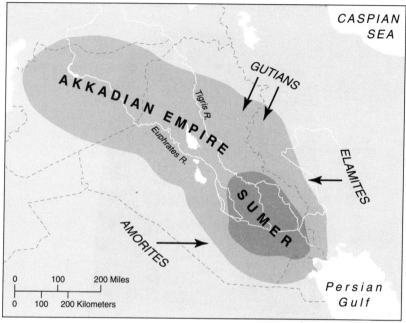

Source: Jon C. Malinowski

Before 2000 B.C. a Semitic people from the west, the Amorites, began to attack parts of the Sumerian Empire. In time, they took control of key Sumerian cities, weakening the empire, which proved more vulnerable when a group called the Elamites invaded and conquered the crucial city of Ur in about 2004 B.C., throwing the region into chaos. City battled city for political supremacy for about two centuries.

Out of the fray emerged the city of Babylon. Culturally, the Babylonians built upon their Sumerian roots, so in many ways earlier cultural traits were perpetuated. The ruler, Hammurabi, conquered and united the cities of the region to create the Babylonian Empire, or Babylonia. Hammurabi is remembered best for the Code of Hammurabi (Figure 6.3), a complex collection of written laws that provided legal protection for nearly all segments of the population, including slaves, women, and children. It also contains laws addressing trade and business disputes. The following passages illustrate the complexity of the Code:

- If a shipbuilder build a boat for some one, and do not make it tight, if during that same year that boat is sent away and suffers injury, the shipbuilder shall take the boat apart and put it togeth-

Figure 6.3 Stele engraved with an image of Hammurabi receiving his legal code from the Sun God.

Source: British Museum 1909.
A Guide to the Babylonian and Assyrian Antiquities, 2nd Edition. London: Harrison and Sons.

er tight at his own expense. The tight boat he shall give to the boat owner.

- If a veterinary surgeon perform a serious operation on an ass or an ox, and cures it, the owner shall pay the surgeon one-sixth of a shekel as a fee.

- If a man knock out the teeth of his equal, his teeth shall be knocked out.

- If a man rent his field for tillage for a fixed rental, and receive the rent of his field, but bad weather come and destroy the harvest, the injury falls upon the tiller of the soil.

- If any one be too lazy to keep his dam in proper condition, and does not so keep it; if then the dam break and all the fields be flooded, then shall he in whose dam the break occurred be sold for money, and the money shall replace the corn which he has caused to be ruined.

This Old Babylonian Empire (Figure 6.4) remained in control until invasions by groups such as the Hittites (from the area of modern Turkey) and the Kassites (possibly from the Zagros of modern Iran) weakened it in the middle of the second millennium B.C. In time, the Kassites came to rule Babylonia and under their rule became one of the most important powers in the region.

Shortly thereafter, the Assyrians, another Semitic-speaking group, became more important in the northern parts of the region. For centuries the Assyrians and Babylonians coexisted, sometimes fighting with each other and sometimes working as allies against other invaders. Struggles with the Assyrians weakened Babylonia to the extent that the region fell under the control of the Elamites, a group from the east. Although Babylon regained some status and dominance during the reign of Nebuchadnezzar I in the twelfth century B.C., constant attacks from groups such as the Aramaeans created centuries of discord.

Beginning in the ninth century B.C., the Chaldeans, a group from near the Persian Gulf, took control of Babylonia and destroyed the remains of the Assyrians. Under the rule of Nebuchadnezzar II, the Neo-Babylonian Empire flourished and expanded to once again control most of Mesopotamia, as shown in Figure 6.4. It is during this time that the famed Hanging Gardens were said to be created. This is also the Biblical period in which Jerusalem is said to have been destroyed, and the Jews sent into captivity in Babylonia.

In 539 B.C. the region fell to Persian (modern Iran) control under the leadership of Cyrus II, marking the end of an independent Babylonia. Although the Mesopotamian region prospered, Persian rule was resented by non-Persian groups. When Alexander the Great arrived with his army in the late fourth century B.C., he was viewed as a liberator, but his plans for the area were never realized due to his early death at age 32. After his death, Greek generals battled among themselves for the territory and accomplished very little in the way of improvements (Metz 1988).

In the mid-third century B.C., a group known as the Parthians took over part of the Mesopotamian region. Their empire stretched from the Euphrates to modern Pakistan. After centuries of rule, the Parthians fell in A.D. 224 to the Sassanids of Persia, who also established a large empire. Centuries of conflict followed for Mesopotamia. In A.D. 641 the Sassanids were defeated by Arab groups who brought with them a fledgling religion, Islam. The Arab period had begun!

MUSLIM ARAB PERIOD

This period established the Islamic faith in Iraq. Despite myths to the contrary, Arabs did not convert individuals at the point of the sword. Many Christians in

Figure 6.4 The Babylonian Empire

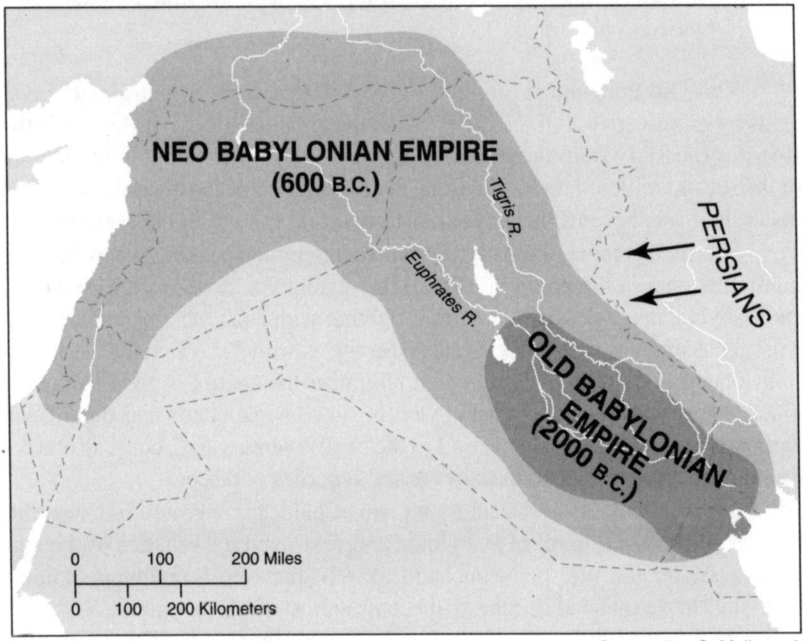

Source: Jon C. Malinowski

the region, for example, continued to practice their religion for centuries after the beginning of Muslim rule, but chose to speak Arabic instead of Aramaic and to pay taxes to the their new rulers. In time, the majority of the population converted to Islam, but for the most part this did not occur immediately or under threat of violence.

At first, what is now Iraq was ruled by the Umayyad caliphs (rulers) based in Damascus, and Arabic gradually replaced Greek and Persian as the vernacular language in the region. In time, Mesopotamia would become the center of the Islamic world under the Abbasid Caliphate that ruled from Baghdad during the period A.D. 750–1258 (see Figure 6.5). Their rule coincided with a period of great intellectual and agricultural development and a flourishing of the young Islamic faith. During this golden age the world was given classic stories such as "Ali Baba and the Forty Thieves," the chronicles of Sinbad, and other tales from the *Arabian Nights*. The society ruled by the Abbasid Caliphate had a mix of Sunni and Shiites from Iraq and Iran. For an extended time there was a balance between these groups, but eventually the death of a caliph caused a succession fight between two sons. The civil war that followed weakened Abbasid rule and precipitated a slow, two-hundred-year decline in the supremacy of Baghdad (Metz, 1988; Polk, 1991).

Thirteenth-century Mongol invasions brought an end to the period of sophisticated urban life enjoyed by the people of this region. The grandson of

Figure 6.5 Territorial Extent of the 5th Century B.C. Persian Empire and Abbasid Caliphate (8th–13th Centuries)

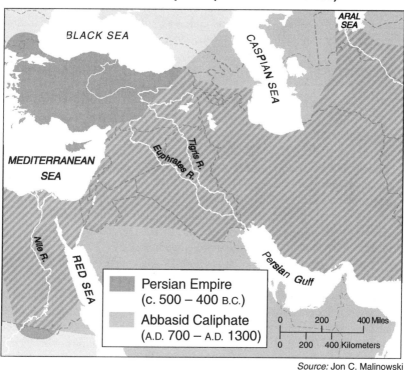

Source: Jon C. Malinowski

Genghis Khan, Hulagu Khan, completed what his grandfather had started in 1227 by extending Mongol rule into Iran and into Iraq. He and later Tamerlane were ruthless invaders, burning, pillaging, and killing to achieve victory. They are responsible for the destruction of many Abbasid advances, including extensive canal and irrigation systems along the Tigris and Euphrates Rivers (Metz, 1988).

THE OTTOMAN ERA (1534–1918)

Between 1258 and 1534, chaos and survival best summarize the conditions in Iraq. A change began with the conquest by the Ottoman Turks of Kurdistan first and then of much of modern Iraq and Iran. The Turks made little effort to change those they conquered, and they generally left existing government systems in control. More commonly, they would merely place a governor in charge of an area and let the locals run the day-to-day operations. This situation worked well for the Sunni population of Iraq, who over time gained administrative experience that contributed to their ability to control the country in the twentieth century. The Shiite population was outside the main political discourse during this period in Iraq, a situation that tended to reduce their access to education and other government-controlled benefits. During the almost 400 years under Ottoman "control," tribal and religious differences caused numerous smaller conflicts. This period ended with the fall of the Ottoman Empire in the early twentieth century (Metz, 1988; Polk, 1991).

THE BRITISH PERIOD AND INDEPENDENT IRAQ

To enlist the support of the local population during World War I (WWI) the British reached agreements that would grant independence to Iraq. Much has been written about the impact of this period and the influence of Britain (and the West) on Iraq (Polk, 1991; Metz, 1988). In short, British agreements had led the way for an independent Iraq. After WWI and a British victory in the region, three Ottoman provinces were combined to form a Class A mandate entrusted to Great Britain under the supervision of the League of Nations. Mandates were a form of indirect rule in which local leaders controlled a country with foreign supervisors overseeing their moves. An important note is that the country created by the British mandate had no history as a unified nation-state.

Facing pressure from Iraqi Arabs seeking quick independence, the British installed King Faisal as head-of-state in a lopsided 1921 election. Granting significant power to local leaders in Kurd and Shiite areas, King Faisal was able to hold power. Geographically, government land reform measures strengthened local areas while modernization created an economic pull to cities, where urban populations grew. By 1932 and official independence, the real power balance in Iraq lay in the urban areas and not with the tribal factions. Urban populations were predominately Sunni and controlled most government offices.

Independence, however, was not entirely peaceful; conflict among the diverse peoples of Iraq challenged the peace. Metz (1988) notes:

> The declaration of statehood and the imposition of fixed boundaries triggered an intense competition for power in the new entity. Sunnis and Shias, cities and tribes, shaykhs and tribesmen, Assyrians and Kurds, pan-Arabists and Iraqi nationalists—all fought vigorously for places in the emerging state structure.

This statement highlights the complex nature of Iraqi culture, the competing factions causing tension, and the centrifugal forces within the new state. One overlooked aspect is the conflict resulting from the sometimes arbitrary delineation of borders by European governments. Some groups had ended up within Iraqi borders against their will. The Kurds of northern Iraq were and are the most notable, finding themselves divided among several countries. Kurdish dislike for Iraq's government would become a wedge that Iran would use decades later to play the Kurds against the Iraqis during their long war in the 1980s.

The early years of modern Iraq also saw the beginning of petroleum exploration and exploitation. The first oil concessions were granted to a foreign company in 1925. In 1931, the Iraq Petroleum Company, comprised of Royal-Dutch Shell, the Anglo-Persian Oil Company, Standard Oil, and others signed an agreement with the Iraqi government to develop oil fields in the area around Mosul in the north. Pipelines were soon completed to the Mediterranean ports of Haifa and Tripoli.

The turmoil in Iraq was exacerbated during WWII when British troops once again returned to Iraq to put down a resistance movement. With the support of the Iraqi monarchy, they occupied large areas of the country, a move that Iraqi nationalists and pan-Arabists would not forget and often used as a reason to oppose the monarchy. After 1945 the country saw a series of short-lived governments that were generally pro-British. Most opposition parties were banned, stunting real democratic reforms. Ethnic tensions continued to brew, putting pressure on the fragile government. This period witnessed a significant rise of pan-Arab nationalism following Nasser's 1952 ascension to power in Egypt, somewhat displacing a more strictly Iraqi nationalism that had grown in previous decades. This period also witnessed an increased importance for the socialist Baath Party, founded in Syria.

Finally, on July 14, 1958, a coup overthrew the monarchy and established a republic under the leadership of 'Abd al-Karim Qasim. Qasim, never really able to establish stability, however, was plagued by disagreements among pan-Arab nationalists, communists, and Kurdish factions. Baathist coups brought new governments in 1963 and 1968. After the 1968 coup, the Baath government simultaneously courted and attacked communist and Kurd political parties. By instituting social and economic reforms, blasting Israeli and foreign presence in the region, and allying itself with the Soviet Union, the Baath government increased its power and public support. Rising oil prices in the 1970s also filled government coffers.

THE HUSSEIN ERA

In 1979, control of the government passed to Saddam Hussein. From a humble background, Hussein had joined the Baath Party during the 1950s. Actively involved in a 1959 assassination attempt on 'Abd al-Karim Qasim, he had to flee the country for several years. Returning in 1963, he rose to a high level in the party, which served him well after the 1968 coup that brought the Baathists to power. Throughout the 1970s he worked on various domestic and foreign issues for the government and party while slowly gathering a close-knit group of powerful supporters, often family members. When the country's leader, President al-Bakr, resigned in 1979, Hussein became Iraq's president and chairman of the Revolutionary Command Council.

The Hussein era, marked by relative internal stability and a sense of national unity, came about through a consolidation of power achieved by reducing internal opposition and by signing a treaty with Iran that ended, for a time, the use of the Kurds to incite dissent within their respective countries. Internally, opposition of any kind was brutally eliminated. Oil revenue continued to be instrumental in the prosperity enjoyed by Iraq prior to the Iran-Iraq war. With the war, started by Iraq, came a desperate time for both countries.

The reasons for the war are reported in both their current and historical context by Polk (1991) in great detail. It bears noting that Saddam Hussein has often been driven by both fear and opportunity. He noted the Shiite-led religious fundamental movement that toppled the shah of Iran and feared similar problems among Iraq's majority Shiite population. This was the first of two strategic miscalculations by Hussein because it led him to fight a protracted land war with Iran in which neither side achieved anything resembling a victory. The second and even more costly error was his decision to attack and occupy Kuwait in 1990. That strategic failure resulted in massive losses at the hands of a coalition force led by the United States. The ramifications and unresolved issues of that conflict led to over a decade of United Nations (UN) economic sanctions, demands to disarm, and the implementation of "no-fly" zones in northern and southern Iraq.

After the terrorist attacks of September 11, 2001, Hussein became a focus of the U.S. campaign against states that support terrorism. Much of the attention centered on whether or not Iraq had truly disarmed as demanded by the UN after the Gulf War. Of particular concern to many countries was the issue of whether Iraq still possessed weapons of mass destruction, such as chemical or biological munitions. It is well documented that Iraq had used chemical weapons against Iraqi citizens and during the Iran-Iraq War. UN inspectors, who had been in Iraq for much of the 1990s, once again entered Iraq after the UN passed another resolution in fall 2002 demanding that Hussein disarm. While Saddam made some concessions to the inspectors, there was deep debate in the international community about whether he was genuinely disarming or simply stalling and deceiving the world.

Despite deep divisions in the international community about whether war was necessary, a coalition led by the United States and Great Britain's troops invaded Iraq in March 2003 to end the Hussein regime.

CONCLUSION

Iraqi history is a complex series of invasions and power changes. Groups entering Iraq from the north, east, south, and west have all controlled the region at one time or another over the past 5,000 years. In addition, Mesopotamian history shows that groups within the region often rose to challenge the central government or ruling power. Perhaps these historical realities partly explain Saddam Hussein's history of mistrust toward 1) his neighbors, a mistrust that led him to war in each of the past two decades; and, 2) minority groups within his own country, such as the Shiite and Kurdish populations. Now, as it has done so many times throughout history, Mesopotamia faces a time of political and social change.

7

Cultural Geography

Jon C. Malinowski

Key Points

- Islam is the dominant religion in Iraq, but there are divisions mainly among Shiite and Sunni groups
- Iraq is the home of many important Shiite holy sites, including Karbala and Najaf
- Arabic is spoken by 80 percent of the population
- Kurdish languages are spoken by nearly 20 percent of Iraqis
- The uniqueness of Kurdish culture goes beyond language differences

---

Statistical tables reveal that the majority of Iraq's population practices the same religion and speaks the same language. But beneath the numbers lies a cultural pattern that is much more diverse and complicated. An understanding of language and religion in Iraq provides a necessary foundation for insight into Iraqi politics, geopolitics, and history.

RELIGION

The power of religion to move people to commit acts of great kindness or horrific violence has been quite evident in recent years. But to truly understand the impact of religion it is important to go beyond simple images of religious zealots protesting or communities banding together to help disaster victims. Religion affects much more than just its followers' spiritual well-being. Indeed, there are few aspects of society that are not influenced by religious ideas. Food taboos affect dietary practices and agricultural practices, worship schedules determine the workweek, and ethical norms underlie legal codes. So when considering Iraq's religious diversity, it is important to remember that the impact of religion might go far beyond religious buildings and basic theological beliefs.

Islam

Nearly 97 percent of the Iraqi population call Islam their religion, but within the Muslim population an important division exists. Islam is divided into two major sects, Sunni and Shiite. This division dates back to the years just after the death of the prophet Muhammad in June of A.D. 632. The religious community that Muhammad had established in present-day Saudi Arabia needed to appoint a successor for their leader in the absence of a male heir. In a space of just 25 years three men succeeded Muhammad as the head of the young religion. The third, Uthman, faced opposition from supporters of Muhammad's son-in-law Ali. When Uthman fell to a murderer in 656, Ali became the head, or caliph, of the religion. After Ali's murder five years later, disagreement arose over whether his successor should be a member of his family or someone chosen from the wider religious community. Muslims known as Shiites today are those who insisted that the successor should be one of Ali's family members. Later history would further divide the Shiites, and today numerous offshoots of Shiite sects can be identified across the Muslim world. In total, Shiites make up only 10–15 percent of all Muslims worldwide. Sunnis, who felt that a qualified Muslim could lead the early community, spread Islamic ideas through expansion and conquest, and now account for 85–90 percent of all Muslims.

Shiites survived as a minority in most areas, but played a critical role during several periods in the history of the region. They were involved in the founding of the Abassid Caliphate (750–945), centered around a grand new city known as Baghdad. Later the Abassid rulers would turn away from Shiite beliefs and violently put down Shiite uprisings. A Shiite state was established in Baghdad in 945 when the Buyids seized control of the city. Their rule lasted only until 1055, when they were overthrown by the Seljuk Turks. A small Shiite dynasty known as the Hamdanids also arose around the same time in Northern Iraq and parts of modern Syria. After the Seljuk Turks consolidated power, Shiite states disappeared until the Safavid Empire in Persia (modern Iran) adopted Shia Islam as the state religion in the sixteenth century. The Persian Safavids at times controlled much of modern Iraq. From this background it is apparent that at various historical moments Shiite influence in Iraq has been important.

The legacy of this history is that Shiites are the majority sect of Islam within Iraq, totaling about 53 percent of the Muslim population. Nearly all Shiite Iraqis are ethnically Arab with a small number of Turkomans, mostly in the north, being the exception. Sunnis account for 42 percent of the Muslim population and include not only Arabs, but also Kurds, Turks, and Turkomans. Because of Iraqi politics, Sunni Muslims are the more powerful group within the country. The Baath Party was dominated by Sunnis and repressed and even attacked Shiite populations. In addition, they promoted a more secular state than other Muslim countries, thus reducing the importance of religion and religious traditions in society.

Geographically, Shiites have traditionally been located in the eastern and southern parts of the country while Sunnis have inhabited central and northern areas (Figure 7.1 [C-9]). Hussein's repression of Shiite "Marsh Arabs" and the draining of their riparian homelands in southern Iraq have reduced their numbers greatly, as tens of thousands fled to Iran for refuge. Hussein had long feared that Shiite populations posed a threat to his rule or would side with Iran during conflicts. For the most part, Shiite populations in Iraq actually supported their country in its long war with Iran during the 1980s.

In addition, Shiite Iraqis are more likely to be rural communities than Sunnis. During the Ottoman rule, the Sunni tradition was emphasized in the schools and preferred in the government, and accordingly, Shiites stayed away from public education and government service. This rural–urban divide remains today.

Another important geographical aspect of Islam in Iraq is the existence of important holy sites (Figures 7.2 and 7.3 [C-10]). The most important is the Shrine to 'Abu 'Abdu'llah Hussein ibn 'Ali, the third imam (religious leader) of Shiite Islam, in Karbala. In A.D. 680 Hussein, the son of Ali, traveled with a small army to Karbala to depose an unjust ruler. His forces were outmatched, and Hussein and his family were brutally killed, ending the hope that Ali's descendants would rule the Muslim world. For modern Shiites, this death is seen as a sacrificial gesture symbolic of the fight against injustice. The religious significance of Karbala is second only to Mecca for Shiites. Many Shiites make pilgrimages to Karbala, and small tablets made from Karbala clay are sent around the world for use during prayer.

A second important holy site is at Najaf, which Shiites believe is the resting place of Ali himself. The city has numerous Shiite seminaries and gave refuge to the Ayatollah Khomeini from 1963 to 1978. Khomeini's expulsion from Najaf by Saddam Hussein, at the request of the shah of Iran, was a source of bitterness among Iran's Shiite community during the Iran-Iraq War. Two other important religious sites are Samarra and Kadhimain, a suburb of Baghdad.

Christianity

Naturally, other religious populations in Iraq are much smaller in number. Christians number about 3–4 percent, but are divided among different sects, such as Assyrians, Chaldean Catholics, Jacobites, and Syrian Catholics. The plight of Christians in Iraq has been difficult in recent years, and many now live either in refugee camps in Jordan or have joined family in Europe or North America.

Followers of the Assyrian Church are modern descendants of the Nestorian Christians of old. Mainstream Christianity considered Nestorian Christianity to be a heresy because of different views on the nature of Christ's humanity and divinity. Their services often use a form of ancient Aramaic, generally recognized to be the language of Christ. Assyrian churches can be found in Iran and Syria, as well as in North America. There is also some connection to Malabar Christians in India. Chaldeans are former Nestorian Christians who have been aligned with the Roman Catholic Church since the sixteenth century.

Jacobites and Syrian Catholics also trace their origins to so-called heretical movements. Again, these ancient Christian groups had different conceptions of Christ's divine nature. Syrian Catholics are Jacobites who are now in communion with the Catholic Church in Rome.

Yazidis

Numbering less than 1,000 worldwide, the Yazidis are a religious group based on a combination of Zoroastrian, Manichean, Christian, Jewish, and Islamic ideas. They are thought to be descended from followers of a seventh-century Islamic ruler, the caliph Yazid. Their society is well-organized and somewhat secretive. Worship revolves around the angel Malak Ta'us, "Peacock Angel," one of seven angels subordinate to a single but absent supreme God. Yazidis can be found near Mosul in Iraq as well as in Iran, Armenia, Turkey, and Syria. Nearly all Yazidis are ethnic Kurds.

Mandaeans and Other Groups

Mandaeans are followers of an old religion with uncertain origins, and they number less than 100,000 in southern Iraq, Iran, Europe, and North America. The religion is Gnostic in its beliefs, holding that esoteric knowledge is critical for salvation. Mandaeans emphasize the importance of communal baptism, and although they are neither Jewish nor Christian, they revere John the Baptist. The community is centered in Baghdad and al Amarah. Other religions in Iraq include small Jewish and Bah'ai communities. Jews in Iraq may have numbered as many as 140,000 after World War II, but many migrated to Israel after its founding in 1948 (Darvish, 1987).

LANGUAGE

According to the Ethnologue database, Iraqis speak 23 native languages (Ethnologue, 2000). But this diverse collection is dominated by various dialects of Arabic, which together is spoken by about 80 percent of the population. In the central part of the country, including the area around Baghdad, Iraqis generally speak Mesopotamian Spoken Arabic (Figure 7.4). In the western parts of the country forms of Najdi Spoken Arabic dominate among Bedouin communities. In the southeast, Gulf Arabic is spoken by a small number of Iraqis.

The second most commonly spoken language is Kurdi, spoken by some members of a Kurdish population that makes up nearly 20 percent of the population. Kurdi speakers probably number around 3 million. Kurdish areas are mostly in the northern and eastern sections of the country. Another Kurdish

Figure 7.4 Major and Minor Languages of Iraq

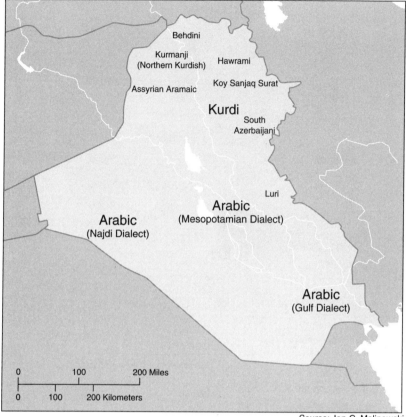

language is Kurmanji, or Northern Kurdish, spoken by a relatively small number of Iraqi Kurds. It should not be assumed that all Kurds speak the same language, or that communication among dialects is easy. Kurdish populations are found Iran, Iraq, Syria, Turkey, and Azerbaijan and from one end of this area to the other significant linguistic differences can be found. The differences among Kurdish speakers have been described as being as distinctive as those between English and German (Ciment, 1996).

Unlike Arabic, which is classified as a Semitic language, Kurdish languages are Indo-European in origin, which connects the ancient history of the Kurds more closely to Iran, where the Indo-European language Farsi is spoken, than to Iraq or Turkey. Other Indo-European languages spoken in Iraq include Farsi, Luri, Hawrami, and Behdini, the first two also being spoken in Iran.

Other notable languages include Assyrian Neo-Aramaic, Chaldean Neo-Aramaic, South Azerbaijani, and Armenian. The diversity of languages in the

Cultural Geography

Figure 7.5 Dominant Kurdish Areas

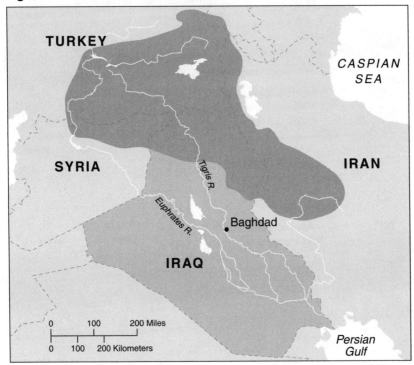

Source: Jon C. Malinowski based on CIA information

mountainous terrain of the north highlights how terrain can provide protection for small ethnic communities, who might get overrun in the accessible river valleys.

THE KURDS

Although Kurdish religion and language have already been addressed, Kurdish culture deserves additional discussion. To review, Kurds are predominantly Sunni Muslims, but some Shiites, Christians, and Yazidis can be found as well. Kurdish languages are distinctive but show great variation throughout the Kurdish areas of Iraq, Iran, Turkey, and Syria, an area sometimes known as "Kurdistan" (see Figure 7.5). There is no single Kurdish language.

In addition to linguistic differentiation, Kurds also have different dress, music, myths, and social customs than Arab Iraqis or their Arab neighbors in Syria, Turkey, or Iran. Ciment (1996) argues that Kurdish culture must be understood in the context of a historical conflict between the mountain-dwelling Kurds and groups from the lowlands. He points to a tradition of highway robbery, thuggery, and violence as evidence of a need to survive tough conditions in a difficult environment. A warrior ethic is also attributed to the Kurds.

Kurdish society in Iraq remains tribal in many, but not all, areas. Households and marriage ties are important to establish lineages, which combine with geographic location and political affiliation to form clans, which may order themselves to form tribes. Tribes are led by chiefs, known as *aghas*, drawn from important families. Chiefs gain much of their authority from sheikhs, important leaders who have a part-religious, part-political authority.

CONCLUSION

Iraq's cultural geography is much more fragmented than many Americans realize. Divisions among Muslim sects, linguistic and religious complexity, and ethnic diversity are all realities of Iraq's historic human landscape. There is no typical Iraqi citizen any more than there is a typical American, and this is an important consideration in any dealings with Iraq.

8

Political Geography

Andrew D. Lohman

Key Points

- Saddam Hussein's quarter-century in power reflected rule through shrewd politics and the ruthless destruction of any opposition
- The two strongest sources of opposition during the Saddam Hussein era were the Kurdish populations in the North and Shiite Muslims in the South
- Although a minority, Sunni Muslims dominated politics when Saddam Hussein was in power
- Kurdish politics has often been characterized by internal disagreements and fighting

The modern state of Iraq occupies the heart of historical Mesopotamia, "the land amidst the rivers," between the Tigris and Euphrates Rivers (de Blij & Muller, 2001). Although there have been countless monarchs, dynasties, and political powers that have ruled over this region throughout its history, its distinction as one of the world's oldest civilizations, and arguably one of the most influential culture hearths in human history, continues to influence the politics of the country and its role in the region. Despite this long and distinctive history, Iraq under Saddam Hussein was widely seen as notorious rather than historically significant because of over a decade of defiance of the global community. This chapter will discuss Iraq's political geography, including its relations with the international community and its internal political dynamics.

THE STATE

Territorially, Iraq encompasses an area of approximately 437,072 sq kms (168,700 sq mi), slightly larger in size than California (U.S. Department of State, 2002). It has a narrow outlet to the Persian Gulf (only 35 miles of coastline), an area that has been periodically fought over with its neighbors Iran and Kuwait. Iraq has over 3,631 kms (2,255 mi) of land boundaries and shares bor-

47

ders with Iran (905 mi/1,458 km), Saudi Arabia (505 mi/814 km), Syria (376 mi/605 km), Turkey (205 mi/331 km) Kuwait (150 mi/242 km), and Jordan (112 mi/181 km) (CIA *World Factbook*, 2002). Iraq has had several boundary disputes with its neighbors along these borders, primarily with Iran over the Shatt al Arab (one of the attributed causes of the Iran-Iraq War), and with Kuwait, which Hussein claimed was Iraq's nineteenth province to justify his 1990 invasion of Kuwait (Anderson, 2000).

Despite cultural, ethnic, and religious ties with the peoples of the neighboring states, Iraq's relations with these countries has been described as tenuous and strained (Held, 2000). Iraq has long "considered itself the eastern anchor of the Arab world and the natural eastern flank of the Arab Middle East," and as such, has tried to assume its perceived position as a leader in the Arab and Islamic world (Held, 2000). Iraq, however, has often been described as a pariah state by countries in and out of the region. Its invasion of Kuwait, defiance of the international community, and actions toward its neighbors both during and after the Gulf War kept tensions in the region high after 1990 (Held, 2000).

Although the ruling regimes in both Syria and Iraq were factions of the Baath Party during the Hussein era, a split between the ideologies of these factions caused a rift between these two states beginning in the early 1960s (Sela, 1998; Held, 2000). Relations with Iran are, of course, still strained after the eight-year war between the two in the 1980s, and Iraq's interaction with Saudi Arabia and Kuwait still reflects animosities from the Gulf War era. Occasionally, tensions flared with Turkey or Syria over water issues, because the headwaters of the Tigris and Euphrates Rivers lie in Turkey and flow through Syria before reaching Iraq. Tensions were often heightened among these three countries, and sometimes between Iran as well, because of Kurdish demands for increased autonomy and outright independence. Although Iraq's shortest boundary is with Jordan, this country has proved to be Iraq's most cordial neighbor and supporter. The Hashemite Kingdom of Jordan has gone to great lengths to maintain open relations with Iraq in the past 20 years and has served as a transit area for moving goods and supplies into and out of Iraq (Held, 2000).

In addition to Jordan's friendly relations with Iraq, the Palestinians, Sudan, and Yemen have been rather ardent and vocal supporters of Iraq (Anderson, 2000). And despite the tremendous condemnation of Iraq on the international scene, both Russia and France argued for easing the UN-imposed sanctions against the country (Anderson, 2000) and vehemently opposed U.S. and U.K. military intervention to disarm Iraq in 2003.

POLITICAL EVOLUTION

Iraq emerged from the remains of the Ottoman Empire after World War I. In dividing up the territory of the empire in 1918, Britain and France had a hand in parceling the territories and created a system of mandates whereby these European states garnered control over portions of the Middle East (Fromkin,

1989). Iraq, as the country exists today, was created from three provinces of the former Ottoman Empire: Mosul in the north, Baghdad in the center, and Basra in the south (Ciment, 1996). Each of these provinces was, and for the most part still is, dominated by distinct cultural groups, drawn along ethnic, linguistic and religious lines, with a Kurdish majority in the north, a Sunni Muslim dominated central region, and Shia Muslims in the southern region.

This political arrangement was created in 1920 and placed under a British mandate. Because of its central location, advantageous position in the river basin, historical dominance in both political and economic arenas, as well as Britain's relationship with the Sunni factions in the region, Baghdad emerged as the center of political power in the country with the Sunnis in control (Ciment, 1996). This new state was initially created as the Kingdom of Iraq, and Emir Faisal ibn Hussein (of the Hashemites) was installed as the king, a reward for his family's cooperation with the British against the Ottoman Turks during World War I (Fromkin, 1989). Intensely proud of its rich history and contributions to the civilized world, the new country did not relish its political domination under British mandate and clamored for independence, which it received in 1932, the first Arab mandate in the region to gain its freedom (Ciment, 1996). However, independence did not free it from British involvement in its internal affairs, driven, of course, by Iraq's rich oil reserves and the control of that industry by the British-dominated Iraqi Petroleum Corporation (IPC).

Emir Faisal died in 1933. Although his heirs nominally ruled until 1958, when his son was finally overthrown in a bloody coup led by General Abd al-Karim Qasim, there were actually over 20 different governments in power between 1948 and 1958, all of which have been described as weak, pro-Western, and unpopular (Ciment, 1996). At first, General Qasim and his leftist Free Officer's Association were extremely popular for a number of reasons, not the least of which was the expulsion of British and Western influence from the country and the nationalization of the oil industry. Popular support quickly eroded, however, and members of the Baath Party deposed Qasim in 1963 (Ciment, 1996). Only in power for a few months, the Baath Party regime was toppled in late 1963 but regained power in another coup in 1968, and from that time until 2003, the Baath Party, under the firm controlling hand of Saddam Hussein since 1979, ruled Iraq.

Although there were other political parties in Saddam's Iraq, the Baath Party (Arab Renaissance Party) was described by the U.S. Department of State (2002) as "the only recognized political party in regime controlled territory." The Baath Party (also described as the Arab Socialist Resurrection Party) is essentially a socialist party, initially formed in Syria in the late 1940s with the goal of molding the Arab world into "a unified democratic socialist Arab nation" (Kurbani, 1995; Metz, 1990). The party's basic principles were pan-Arab in scope, and along with that, it sought to end the colonial relationships that had plagued the region (Hiro, 1998). While some of the basic Baath ideals changed

Figure 8.1 Political Provinces of Iraq

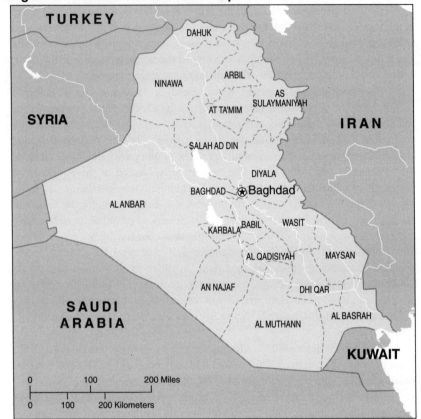

Source: *World Geographical Encyclopedia, Volume 3, Asia*, 1995

under Saddam Hussein's leadership, many assessments contend he still had illusions of pursuing this pan-Arab dream (Kurbani, 1995).

Iraq under Hussein was officially a republic with political rule vested in a nine-member Revolutionary Command Council (RCC) along with a legislative branch of a 250-member National Assembly (CIA *World Factbook*, 2001). All real power in the state, however, was controlled by Saddam Hussein, who held the positions of president, RCC chairman, prime minister, and secretary-general of the Baath Party (U.S. Department of State, 2002). Through control of the Baath Party apparatus, which Saddam filled with his closest friends and relatives, he was able to maintain tight control over the state.

According to Metz, (1990), the Baath Party was essentially an elitist organization with only a small number of full party members (about 30,000) supported by about 10 percent of the population. Political success in Iraq was, of course, dependent entirely on party membership and loyalty to Saddam. Inter-

Political Geography

Table 8.1 Provinces of Iraq

Province	Area (mi^2)	Provincial Capital
Al Anbar	53,461	Ar Ramadi
Al Basrah	7,361	Al Basrah
Al Muthanna	19, 972	As Samawah
Al Qadisiyah	3,147	Ad Diwaniyah
An Najaf	11,126	An Najaf
Arbil*	5,586	Arbil
As Sulaimaniya*	6,571	As Sulaimaniya
At Ta'mim	3,969	Karkuk
Babil	2,497	Al Hillah
Baghdad	283	Baghdad
Dahuk*	2,522	Dihok
Dhi Qar	4,979	An Nasiriyah
Diyala	7,363	Ba'qubah
Karbala[1]	1,943	Al Hindiyah
Maysan	6,204	Al Amarah
Ninawa	14,407	Mosul
Salah ad Din	9,554	Samarra
Wasit	6,621	Al Kut

*Designated as Kurdish Autonomous Region

Source: *World Geographical Encyclopedia, Volume 3, Asia*, 1995

nally, Iraq was divided into eighteen provinces (See Figure 8.1 and Table 8.1), under the administration of a governor (appointed by the president) with what the U.S. Department of State (2002) termed "extensive administrative powers." These provinces were further subdivided into districts and subdistricts and many of these officials, including mayors of cities and many towns were also appointed by the president (Metz, 1990).

In addition to the politically appointed governors, mayors, and other local leaders, the Baath Party organization also played a vital role in administering the state internally, which Metz (1990) described in the following manner in *Iraq: A Country Study*:

> The basic organizational unit of the Baath was the party cell or circle (*halaqah*). Composed of between three and seven members, cells functioned at the neighborhood or the village level, where members met to discuss and to carry out party directives. A minimum of two and a maximum of seven cells formed a party division (*firqah*). Divisions operated in urban quarters, larger villages, offices, factories, schools, and other organizations. Division units were spread throughout the bureaucracy and the military, where they functioned as the ears and eyes of the party. Two to five divisions formed a section (*shabah*). A section operated at the level of a large city quarter, a town, or a rural district. Above the section was the

branch (*fira*), which was composed of at least two sections and which op-
erated at the provincial level. There were twenty-one Baath Party branches
in Iraq, one in each of the eighteen provinces and three in Baghdad. The
union of all the branches formed the party's congress, which elected the
Regional Command. (Metz, 1990)

Saddam Hussein was able to maintain a tight grip on the reins of power in Iraq
by controlling the Baath Party, and by exploiting factions and political allegiances in the country allied along cultural, ethnic, and religious lines. He
learned to do this to his advantage through experience gained in almost 25
years of rule. Alliances between the regime and various tribal groups were the
dominant trend in Iraqi politics (Sela, 1998). Saddam secured allegiances and
loyalties from tribal leaders through both the promise of rewards and by the
threat and use of punishment. The judicial system in Iraq, as described by
the U.S. State Department, included three types of courts: civil (based on the
French model), religious (based on *sharia* or Islamic law), and special courts
(that "try broadly defined national security cases"). With regard to rewards, in
many cases Saddam gave "lands, funds, judicial autonomy over tribal matters,
and exemptions from military services" to buy loyalties from various tribal
leaders (Sela, 1998). His power was further enhanced through manipulating
family bonds and in many cases, "key positions in the regime and control of
major sectors of the economy (were) linked to intermarriages with Saddam
Hussein and ties formed in his village of origin, al-Tikrit" (Eickelman &
Piscatori, 1996).

Although the Sunni-dominated central region of the country was, for the
most part, considered loyal to the regime, Saddam's tinkering with Iraqi politics changed the internal political dynamics. He continued to consolidate and
centralize his power and furthered rampant nepotism as he strengthened the positions of his sons and closest family members. As a result, tribes formerly
loyal to Saddam's regime began to resent their weakening position and influence within domestic politics (Sela, 1998). Dissatisfied with their real or perceived loss of power in domestic affairs, several tribes rebelled against
Saddam, most notably the Jubayr tribe in 1990 and Dulaym tribe in 1995 (Sela,
1998; Ciment, 1996). Two of Saddam's sons-in-law defected to Jordan in
1995, further indicators of discontent within the regime during the last decade
of Hussein's rule. As in all cases of disloyalty, however, those involved in these
breaches of loyalty were dealt with quickly and severely.

Despite reports that other political parties were outlawed, several other
political parties did exist in the country to represent the various ethnic factions
in Iraqi society, but, as mentioned above, they exercised little or no power at
all in government affairs. Of the various groups, tribes, and parties in Iraq, the
Kurds in the north and the Shiites in the south were the most significant.

THE KURDS

The Kurds have received a great deal of attention in the political arena due to their plight as perhaps the most populous stateless nation in the world today. (A stateless nation is a culturally homogeneous group of people who lack the territorial means to achieve statehood [de Blij & Muller, 2001].) Aspiring to statehood, the Kurdish people, in fact, extend across borders and inhabit portions of Syria, Iran, and Turkey as well as Iraq. In their quest for political sovereignty and statehood, the Kurds have, by all accounts, been "at war" with the governments of these countries, and in Iraq the Kurds effectively fought against the Iraqi regime almost continuously for 30 years (Sela, 1998). Although the Baath Party was the only authorized political party in Iraq, the Kurds represented the next most politically powerfully group in Saddam's Iraq and will continue to be important in a post-Hussein Iraq.

In the last two major external conflicts (against Iran between 1980 and 1988 and against the Coalition forces during the 1991 Gulf War), the Kurds tried to take advantage of the opportunity to rebel in the hopes of attaining the territory to create an independent Kurdish state. In both of these attempted rebellions, Iran and the United States reportedly aided the Kurds, but withdrew support when the conflicts with Iraq ended (Ciment, 1996; Sela 1998). Following the cease-fire agreement with Iran in 1988, the Iraqi government subsequently launched a campaign to punish the Kurds for their rebellious acts and support for Iran during the war (Ciment, 1996). In the ensuing Anfal campaign (*Anfal* is the Koranic word for "spoils of war") the Iraqi government attempted to cleanse the Kurdish homeland, so-called Kurdistan, and effect the "removal of hundreds of thousands of Kurds from their mountain fastness and destruction of strategic swaths of their homeland" (Ciment, 1996).

During this two-year operation (1988–1990), Saddam's forces conducted what can be described as a scorched earth campaign. Males of fighting age in many cases were systematically killed, women tortured through starvation and thirst, beaten, raped, and relocated as their homes were destroyed, wells plugged, and fields salted in attempts to make their traditional homelands unlivable (Ciment, 1996). One of the more startling and shocking accusations during this period, though, is reports of Iraqi use of chemical weapons and agents against the Kurds. In one such chemical attack, on the Kurdish city of Halabja in April 1988, there were reportedly between 5,000 and 6,000 Kurdish fatalities (Ciment, 1996; Sela, 1998). While there are numerous reports of such atrocities, many of the Kurds were resettled in Arab cities throughout Iraq, and into government-built housing areas, which have been described as almost being internment camps; they are surrounded by barriers and armed guards (Ciment, 1996).

Despite this treatment at the hands of the Iraqi government, the Kurds again tried to rebel when Iraq was faced with another external conflict during the Gulf War. There are a number of accounts (Ciment, 1996; Sela, 1998) which claim that the United States urged the Kurds to rebel after the Gulf War but failed to provide support when Iraqi ground forces turned on them. As a result, thousands were killed and perhaps as many as one million fled as refugees into Turkey and Iran (Ciment, 1996).

Undoubtedly, such harsh and repressive methods against the Kurds were intended "to break the back of Kurdish resistance and prevent future uprisings" (Ciment, 1996). Furthermore, their resettlement into controlled housing areas and into Arab-dominated cities and regions was intended to disperse them throughout the loyal portions of the population and dilute their political power and ability to unify and renew their efforts.

Although the United Nations was not able to stop Iraqi abuses against the Kurds entirely, the UN took several steps to provide the Kurds a measure of protection. The northern three provinces in Iraq (Arbil, Sulaimaniyah, and Dahuk) were collectively organized into an Autonomous Region in 1992 by order of UN Security Council Resolution 688 (Sela, 1998). These areas, often referred to as safe havens, were created to protect the Kurds from Iraqi reprisals following their rebellion and to create a safe area for Kurdish refugees to return from Turkey and Iran (Sela 1998). Furthermore, the United States sponsored a "no-fly" zone (Figure 8.2) north of the 36th parallel to prohibit Iraqi air power from harassing and bombing these Kurdish regions. Although the no-fly zone prevented Iraqi air power from being used against the Kurds, Iraqi ground forces made many incursions into this region to control parts of it.

Despite the creation of this Autonomous Region in Iraq, political and cultural infighting kept the Kurds from becoming a serious threat to the Iraqi regime. While numerous smaller Kurdish political parties exist, the two dominant parties are the Kurdish Democratic Party (KDP), located primarily in the northwestern part of Kurdistan along the Turkish and Syrian borders, and the Patriotic Union of Kurdistan (PUK), located on the eastern portion of Kurdistan along the Iranian border, with the city of Sulaimaniya serving as its "capital" (Ciment, 1996). Despite their mutual nationalist goal of attaining an independent and sovereign Kurdistan, these Kurdish political parties have been engaged in their own civil war on a relatively regular basis over the last 30 to 40 years. Each of them, including the smaller peripheral parties, have their own traditional militias, known as *Peshmerga* (literally "those who stare death in the face"), which serve as the military arm of their organizations to help achieve their political ends (Ciment, 1996; Sela, 1998). These groups eagerly joined with Coalition forces when fighting broke out in Iraq during March 2003.

Although there have been temporary alliances of Kurdish political parties in the past, such as the Iraqi Kurd Front (IKF), such unions have invariably failed to produce any meaningful coalitions. Kurdish politics, like all political forces in Iraq, are driven by tribal ties and affiliations formed under *aghas*, traditional tribal leaders (Eickelman & Piscatori, 1996). Past attempts to foster effective Kurdish unity, including ties with Kurdish parties and tribes in neighboring Iran, Syria, and Turkey, have been derailed as tribal feuds and

Figure 8.2 The No-Fly Zone

Source: United States Air Force

competition for overall control have pitted Kurd against Kurd (Sela, 1998; Ciment, 1996). Political relationships are often complicated even further as Kurdish factions have been accused of conspiring with Saddam's regime, as well as the governments of neighboring states, against their fellow Kurds in desperate gambles to consolidate their power and gain exclusive control over Kurdistan. In a post-Saddam Iraq, the status of Kurdish areas will again be an important issue for a new government.

THE SHIAS

Estimated at anywhere between 40 to 60 percent of the Iraqi population, the Shias are a potentially powerful political force in the country. Like the Kurds, they rebelled against the (Sunni) Hussein regime numerous times in the past. The last major uprising occurred in 1991 following the end of the Gulf War and ended with Saddam crushing all opposition after Allied troops left the country. Many of the Shias were driven into neighboring Iran or sought shelter in the marsh region among the Madan (the "Marsh Arabs" who inhabit the marshy regions in the lower courses of the Tigris-Euphrates valley regions) (Held, 2000).

In similar fashion to dealing with the Kurds, Saddam's methods for punishing the Shias were swift and ruthless. In an effort to drain the marshy central region between the two rivers, the Iraqi government began construction of the Main Outfall Drain (MOD) in 1992. Although ostensibly constructed to reclaim the land for development in this region, many theorize that this project was merely an effort to destroy and eliminate this sanctuary for the Shias (Held, 2000). This project, along with Saddam's crackdown, forced untold thousands of Shiites to flee to other parts of the country or to Iran, where they hoped to find relative peace from persecution. As further punishment, Held (2000) argues the Iraqi regime was purposely slow in repairing the infrastructure in the Basra region devastated during the Gulf War. In an effort to protect the Shias from Saddam's reprisals, the UN created a southern "no-fly" zone south along the 33rd parallel (Figure 8.2).

CONCLUSION

Obviously the political future of Iraq is quite uncertain. There is little doubt, however, that some of geographic realities of the country, namely regional ethnic groups with political identities, will continue to be an important factor in political stability. If Shia, Sunni, and Kurdish factions can work together, Iraq's future will be more stable. If they fight with each other, or within their own communities, as all have done in the past, Iraq may face years or decades of chaos.

Iraq: A Geographic Portrait

Source: U.S. Marine Corps/Joseph R. Chenelly

Figure 1.1   The Scope of Regional Geography

Source: Adapted from deBlij & Muller, 2001

Figure 2.2 Iraq

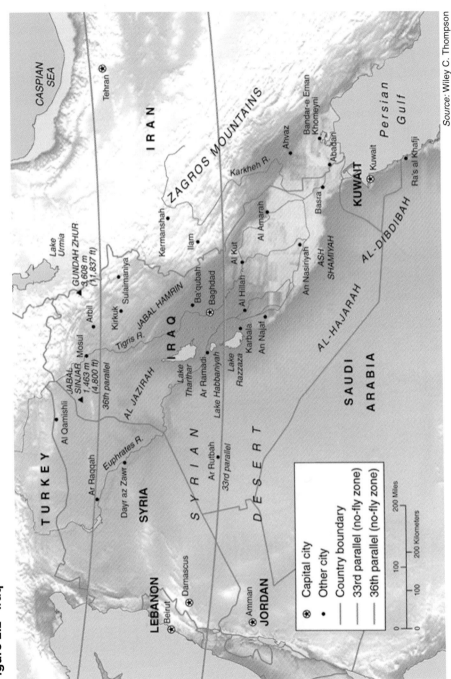

Source: Wiley C. Thompson

Figure 2.3 Iraq in Southwest Asia

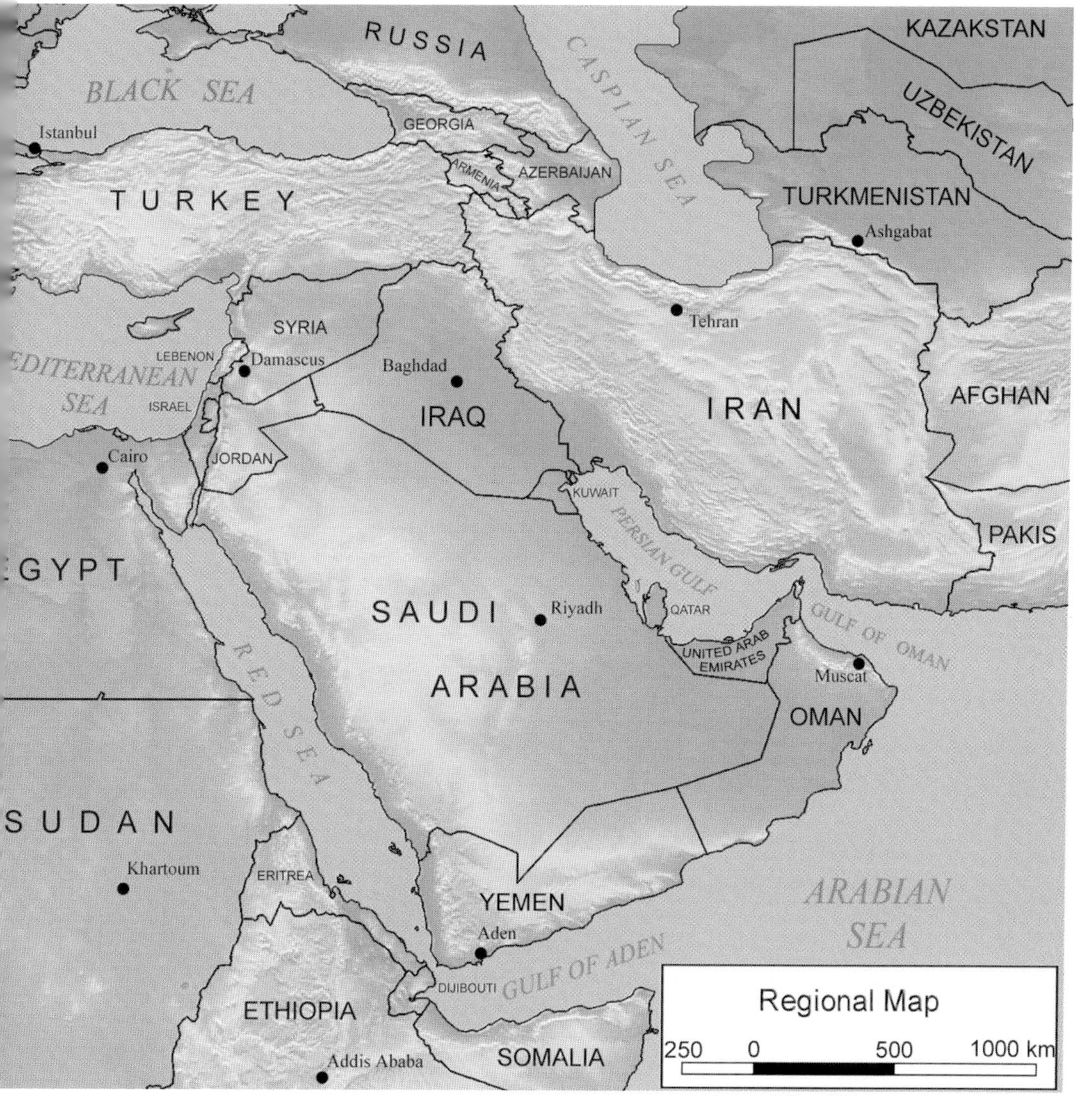

Source: Wiley C.Thompson

Figure 2.5   Satellite Image of Mesopotamia

Source: NASA

This satellite photo of eastern and southern Iraq highlights many aspects of the country's geography. First, note the wide extent of brown desert landscapes, especially in the south and southwest. Second, the influence of the Tigris and Euphrates rivers can clearly be seen from the middle of the country to the Arabian Gulf in the south. The darker sections between the rivers are Iraq's primary agricultural and population centers. Baghdad is east of the two large, black water bodies in the middle of the image. Finally, note the change in color just east of the border with Iran. These are the mighty Zagros Mountains that form a natural boundary along much of Iraq's eastern border.

Figure 3.1 Physical Features of Iraq

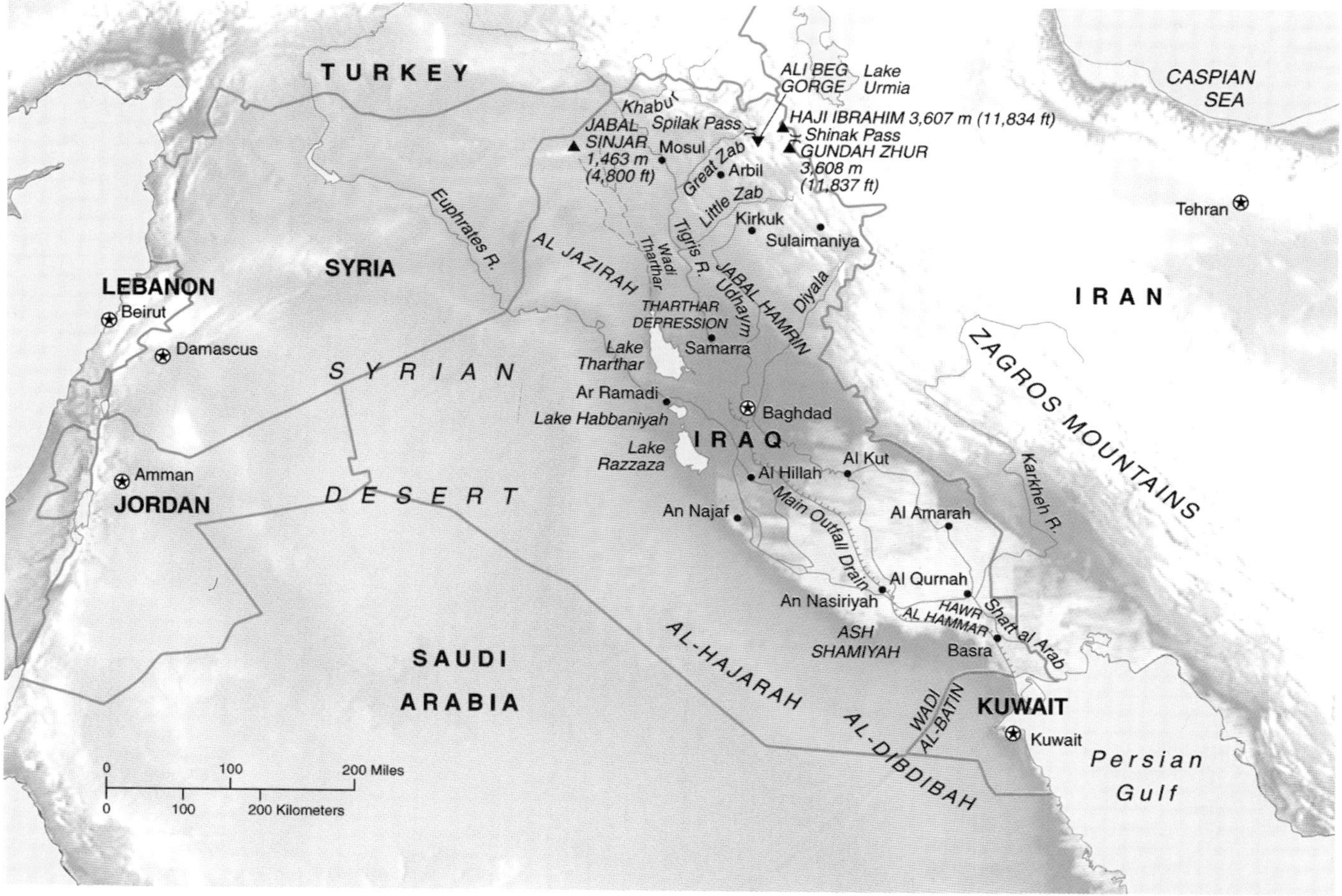

Source: CIA, 2003

Figure 3.2   Hydrology in Southeastern Iraq, June 1994

Former Marshes and Water Diversion Projects in Southeastern Iraq, June 1994

Area of substantial drying
Marsh
Flooded area
Existing canal
Road

Scale 1:2,100,000

0   10   20   30   40 Kilometers
0   10   20   30   40 Miles

Source: CIA, 1994

Figure 3.3 Marsh Decline in Southern Iraq, 1973 & 2000

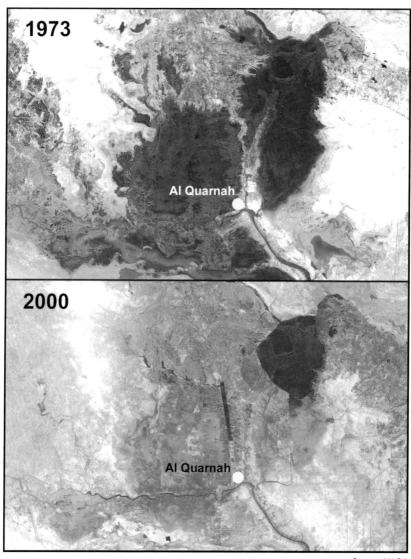

Source: NASA

These false-color, NASA Landsat satellite images of southern Iraq show the extent of marsh decline near the city of Basra. Marsh areas show up as dark red because of dense vegetation, such as marsh grass. Irrigated agricultural areas show up as a bright red. Temporary lakes are light blue and permanent water is almost black. It is clear that between 1973 and 2000 a significant area of marshlands, as much as 85 percent, has been lost. Diversion of river flow to irrigation has reduced the amount of water reaching the area, destroying the lifestyle and culture of the "Marsh Arabs," who have lived in this region for thousands of years. It is believed that Saddam Hussein purposely diverted water to destroy the culture of the area because the Marsh Arabs are Shiite Muslims who typically resisted the Baghdad regime.

Figure 4.5

Source: U.S. Department of Defense

The ferocity of sandstorms in Iraq became clear to many in the West in March 2003 during the early days of U.S.- and British-led operations in Iraq

Figure 9.5

Source: U.S. Department of Defense

The Port of Umm Qasr with HMS *Sir Galahad* in March 2003

Figure 7.1 Ethnolinguistic Groups in Iraq

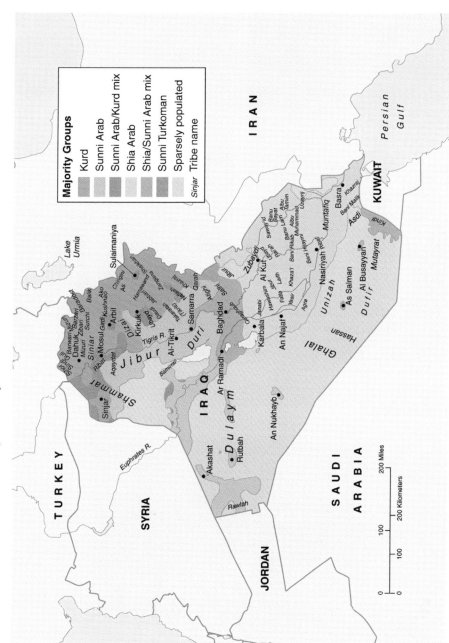

Source: CIA

Figure 7.2 Tomb of Hussein, Karbala

Figure 7.3 Tomb of Ali, Najaf

Source: Copyright (c) 1993 Corel Corporation

Source: Copyright (c) 1993 Corel Corporation

Figure 8.3

Source: U.S. Department of Defense

Figure 9.1 Economic Activity

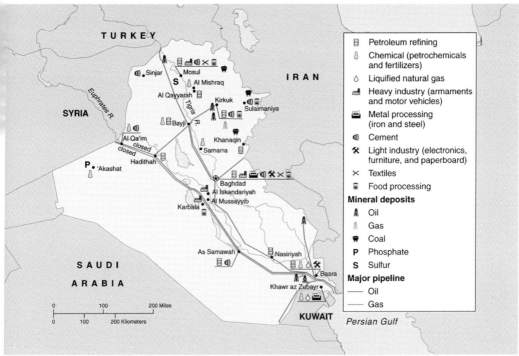

Source: CIA Atlas of the Middle East, 1993

Figure 9.3   Iraq's Oil Infrastructure

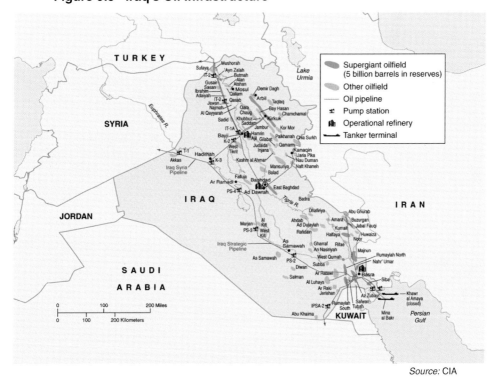

Source: CIA

Figure 9.4 Land Use in Iraq

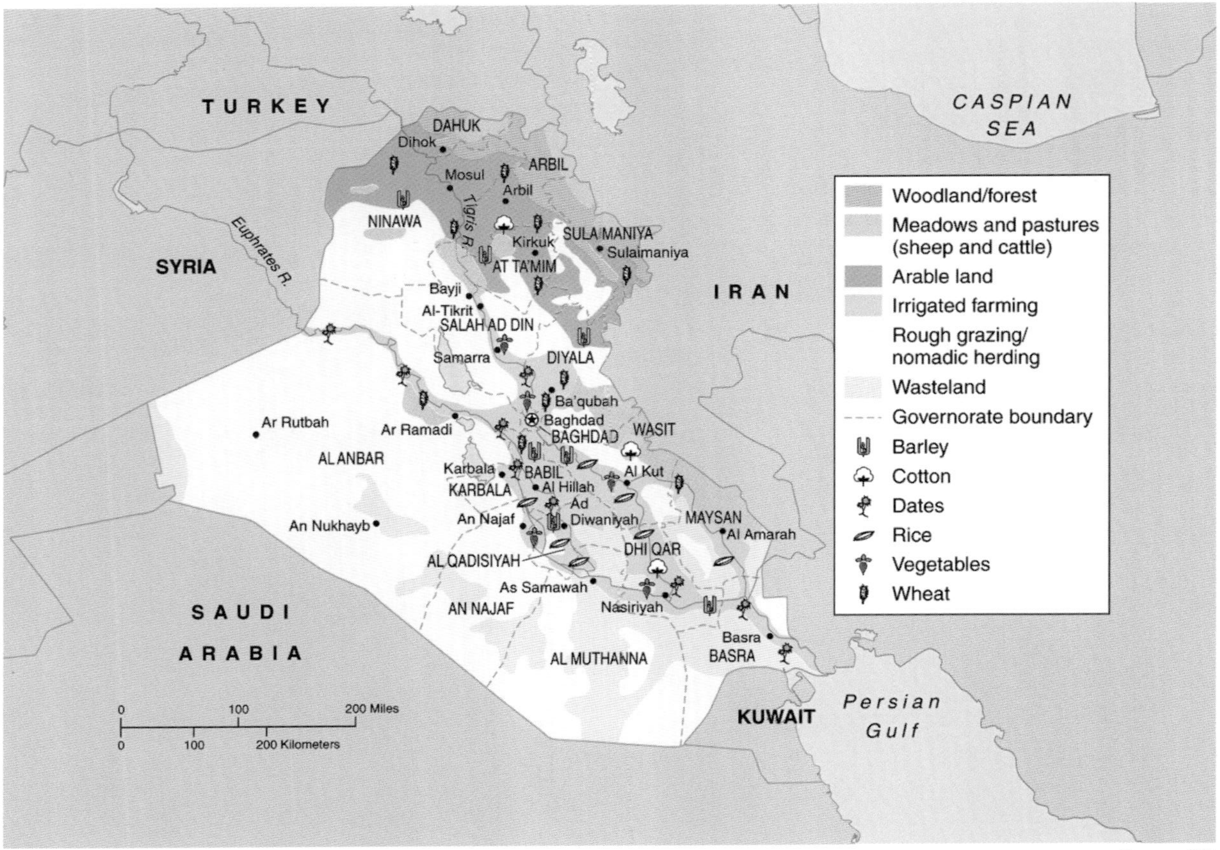

Source: CIA

Figure 10.1 Population Density

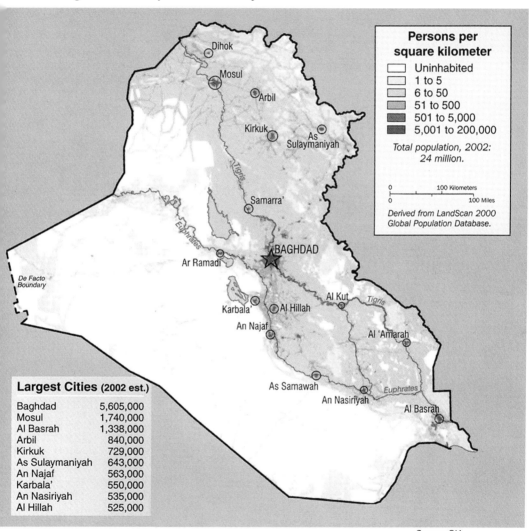

Persons per square kilometer	
☐	Uninhabited
☐	1 to 5
☐	6 to 50
▨	51 to 500
▩	501 to 5,000
▰	5,001 to 200,000

Total population, 2002: 24 million.

Derived from LandScan 2000 Global Population Database.

Largest Cities (2002 est.)	
Baghdad	5,605,000
Mosul	1,740,000
Al Basrah	1,338,000
Arbil	840,000
Kirkuk	729,000
As Sulaymaniyah	643,000
An Najaf	563,000
Karbala'	550,000
An Nasiriyah	535,000
Al Hillah	525,000

Source: CIA

Figure 10.2 Baghdad and Vicinity

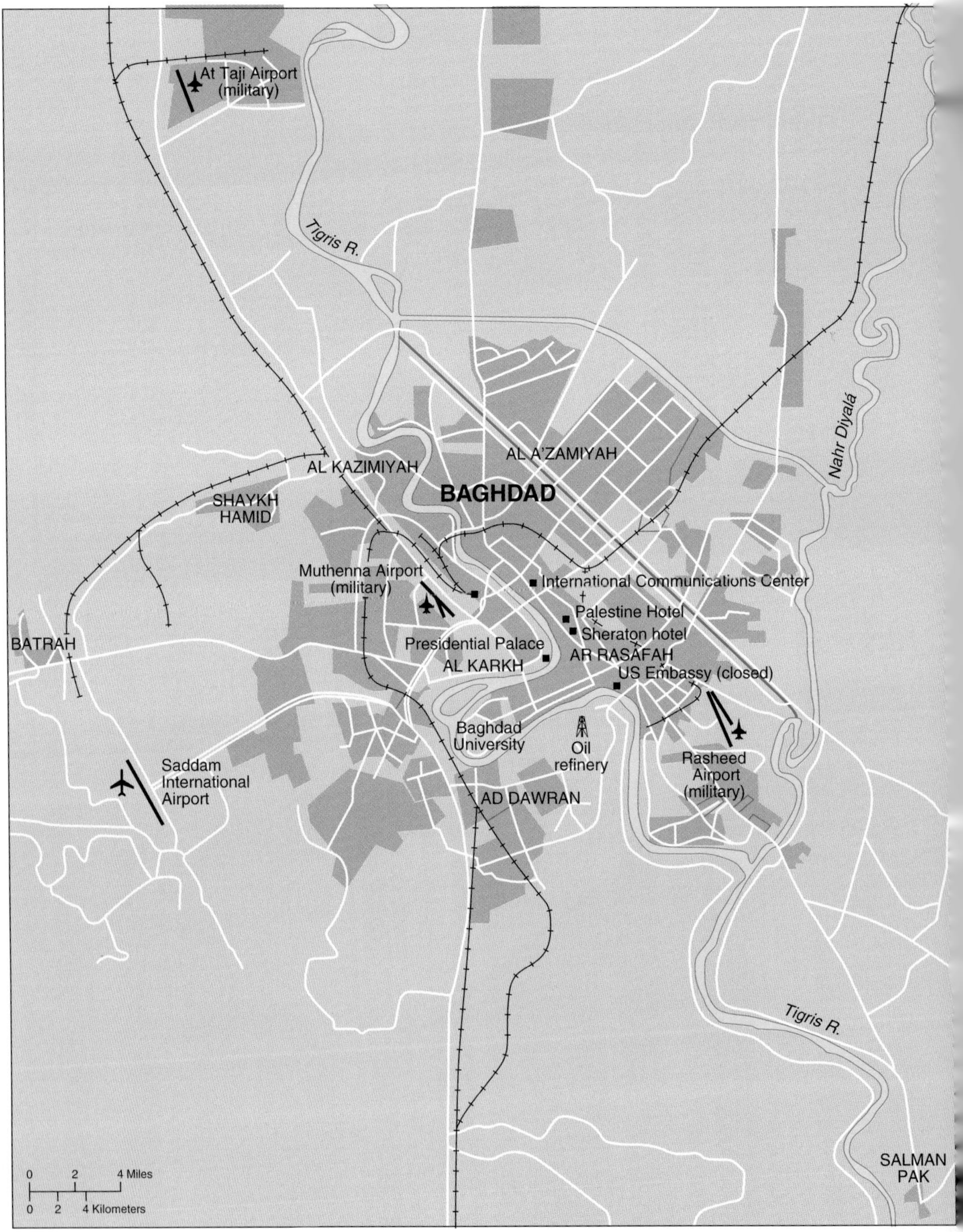

- At Taji Airport (military)
- Tigris R.
- Nahr Diyalá
- AL KAZIMIYAH
- AL A'ZAMIYAH
- **BAGHDAD**
- SHAYKH HAMID
- Muthenna Airport (military)
- International Communications Center
- Palestine Hotel
- Sheraton hotel
- BATRAH
- Presidential Palace
- AR RASAFAH
- AL KARKH
- US Embassy (closed)
- Baghdad University
- Oil refinery
- Rasheed Airport (military)
- Saddam International Airport
- AD DAWRAN
- Tigris R.
- SALMAN PAK

0   2   4 Miles
0   2   4 Kilometers

Source: Defense Mapping Agency. 1991

Figure 10.3 Mosul and Vicinity

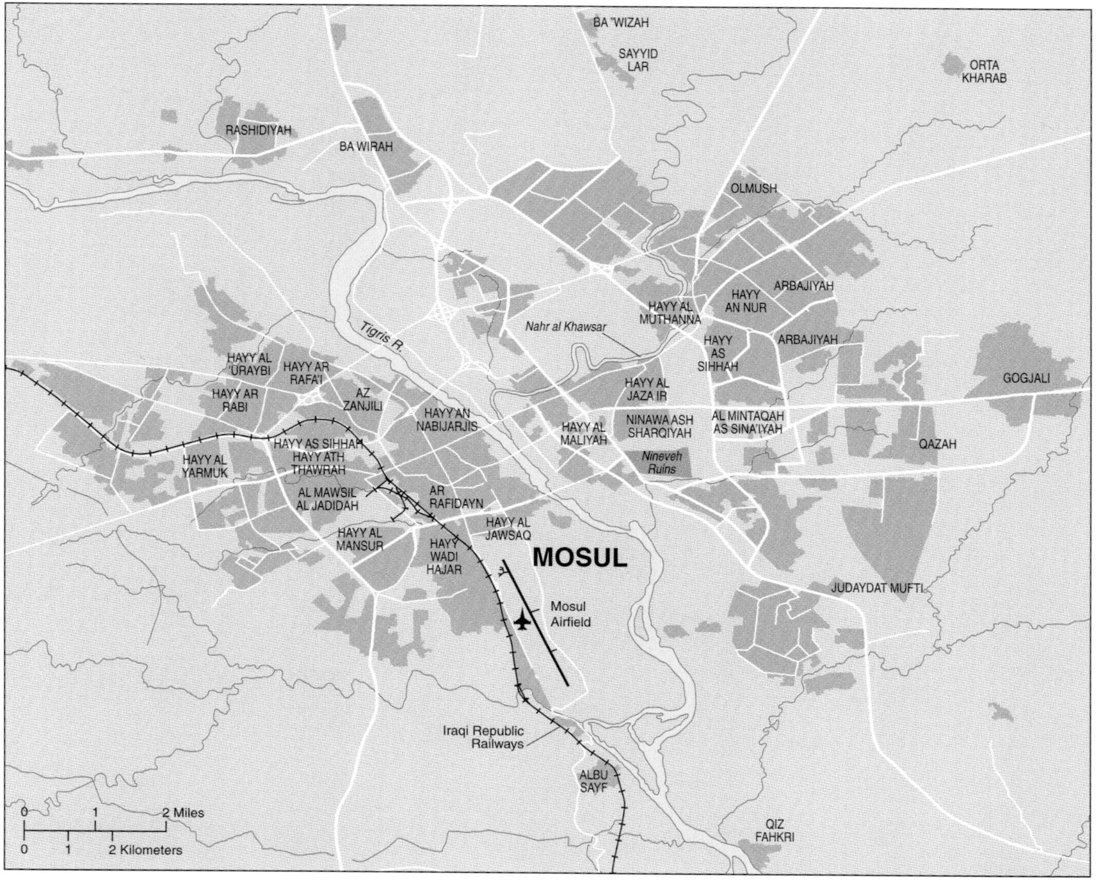

BA'WIZAH

ORTA KHARAB

SAYID LAR

OLMUSH

ARBAJIYAH

HAYY AN NUR

HAYY AL MUTHANNA

ARBAJIYAH

HAYY AS SIHHAH

HAYY AL JAZA'IR

AL MINTAQAH AS SINA'IYAH

HAYY AL MALIYAH

NINAWA ASH SHARQIYAH

Nineveh Ruins

GOGJALI

QAZAH

JUDAYDAT MUFTI

QIZ FAHKRI

BA'WIRAH

Nahr al Khawsar

RASHIDIYAH

HAYY AN NABUAR'US

AR RAFIDAYN

HAYY AL JAWSAQ

MOSUL

Mosul Airfield

ALBU SAYF

Iraqi Republic Railways

Tigris R.

AZ ZANJILI

HAYY AS SIHHAH

HAYY ATH THAWRAH

AL MAWSIL AL JADIDAH

HAYY AL MANSUR

HAYY WADI HAJAR

HAYY AR RAFAI

HAYY AR RABI

HAYY AL URAYBI

HAYY AL YARMUK

0   1   2 Miles

0   1   2 Kilometers

Source: Defense Mapping Agency, 1989

Figure 10.4 Basra and Vicinity

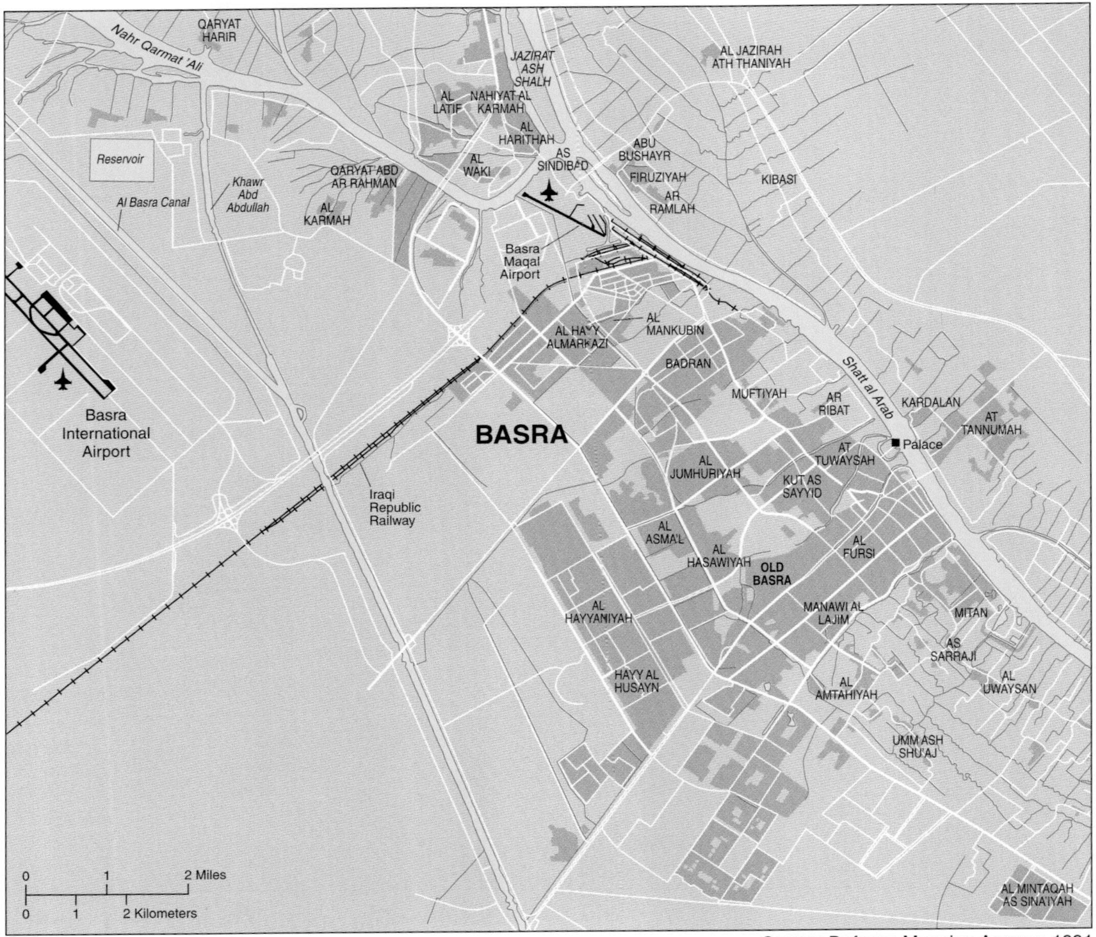

QARYAT HARIR

Nahr Qarmat Ali

JAZIRAT ASH SHALH

AL NAHYAT AL LATIF KARMAH

AL HARITHAH

AS SINDBED

AL WAKI

QARYAT ABD AR RAHMAN

AL KARMAH

Khawr Abd Abdullah

Al Basra Canal

Reservoir

Basra International Airport

Iraqi Republic Railway

Basra Maqal Airport

AL JAZIRAH ATH THANIYAH

KIBASI

ABU BUSHAYR

AR RAMLAH

FIRUZIYAH

AL HAY'Y AL MARKAZI

AL MANKUBIN

BADRAN

MUFTIYAH

Shatt al Arab

AR RIBAT

KARDALAN

AT TANNUMAH

Palace

AT TUWAYSAH

KUT AS SAYYID

AL JUMHURIYAH

AL ASMA'IL

AL HASAWIYAH

OLD BASRA

AL FURSI

MANAWI AL LAJIM

MITAN

AS SARRAJI

UMM ASH SHUJA'I

AL UWAYSAN

AL AMTAHIYAH

AL MINTAQAH AS SINAYAH

BASRA

AL HAYYANIYAH

HAY'A AL HUSAYN

0   1   2 Miles

0   1   2 Kilometers

Source: Defense Mapping Agency, 1991

9

Economic Geography

Albert A. Lahood

Key Points

- Based solely on oil reserves, Iraq is one of the richest countries in the world
- Because of United Nations sanctions and governmental neglect, the Iraqi people are the poorest in the region
- Iraq has almost no capacity to grow its own food

Iraq's economy has been dominated by the oil sector, which has traditionally provided about 95 percent of foreign exchange earnings ... [however] Iraq's economy has been severely damaged by war, mismanagement, corruption and by the sanctions imposed upon it as a result of Iraq's attack on Kuwait in 1990. (CountryWatch, 2001)

Conducting an economic assessment of Iraq is difficult at best. The scant economic data that are available only serve to obscure the fact that since 1995 the Iraqi economy was solely based on the UN Oil-for-Food Program under UN Security Council Resolution 986 (UNSCR 986), and to some unknown extent, illegal oil smuggling operations. Therefore, the health of the Iraqi economy was largely controlled by the UN Security Council through the auspices of the 191 member-states of the UN. As Iraq looks forward in the wake of the spring 2003 conflict, it possesses the resource potential to become one of the region's leading economic powers and top global petroleum producer. The damage to the state's infrastructure resulting from 20-plus years of conflict and neglect must be repaired, however, before Iraq can effectively capitalize on its bountiful resource base.

This chapter will first consider the economic conditions of the Iraqi state on the verge of the Coalition attack in 2003 and its future economic potential. The starting point will be a review of Iraq's primary resource base, and the industrial, agricultural, and transportation sectors. Secondly, the current economic conditions will be explored in terms of how they affect the Iraqi people and their everyday lives.

PRIMARY RESOURCES

Iraq is one of the most resource-rich Middle East and North Africa (MENA) states. In terms of hydrocarbon fuels Iraq has a proven oil reserve of 112 billion barrels and an estimated 215 billion barrels of probable oil in the currently exploited geomorphic formations (Energy Information Administration [EIA], 2001). The oil is primarily concentrated in two areas, Kirkuk in the north and Basra in the southeast near the confluence of the Tigris and Euphrates Rivers (see Figure 9.1 [C-11]).

This establishes Iraq as having the second-largest oil reserve in the world following Saudi Arabia (EIA, 2001). Furthermore, this estimate is understated because Iraq is believed to possess "deeper oil-bearing formations located mainly in the Western Desert region," though this area has yet to be explored (EIA, 2001). This unexplored region could place Iraq in the position of having the world's largest petroleum reserves. Additionally, Iraq has a vast amount of associated and nonassociated natural gas. It has a proven store of 110 trillion cubic feet (Tcf) and 150 Tcf of probable reserves (EIA, 2001). About 70 percent of this natural gas is associated with the oil fields near Kirkuk and Basra (EIA, 2001).

Not only is Iraq well endowed with hydrocarbon fuels, but it also has an abundance of the most important Middle Eastern resource of all, water. Iraq's borders contain the Tigris and Euphrates Rivers, which combined have a flow of 75–85 billion cubic meters annually (Anderson, 2000). This large perennial surface flow makes it one of the top five among the MENA states, and coupled with Iraq's relatively small population, 23.6 million, one of the highest water volumes per capita in the region (CIA, 2001). Iraq has a large amount of surface water crossing its borders, but does not control the headwaters of rivers such as the Tigris and Euphrates. Turkey now effectively controls the flow of both rivers through Iraq because of its Grand Anatolia Project (GAP), which includes a series of large dams and reservoirs (BBC News, 2000).

INDUSTRY

Since the 1950s Iraq's industrial sector has concentrated on the extraction of crude oil and natural gas for export. This sector virtually replaced the agricultural sector that dominated Iraq's economy for thousands of years. The once-private oil industry was nationalized after the shah of Iran's fall so that the Oil Ministry directly oversaw the Iraq National Oil Company.

Iraq has a limited oil refinery industry, capable of producing 400,000 barrels per day (bbl/d) of finished petroleum products (EIA, 2001). Out of Iraq's ten refineries, Baiji North and the Basra, Daura, and Nasiriyay plants in the south are operating around 100,000 bbl/d. Because of the small quantities, it is mainly refined for domestic consumption.

Figure 9.2 Oil Production and Consumption, 1980–2000

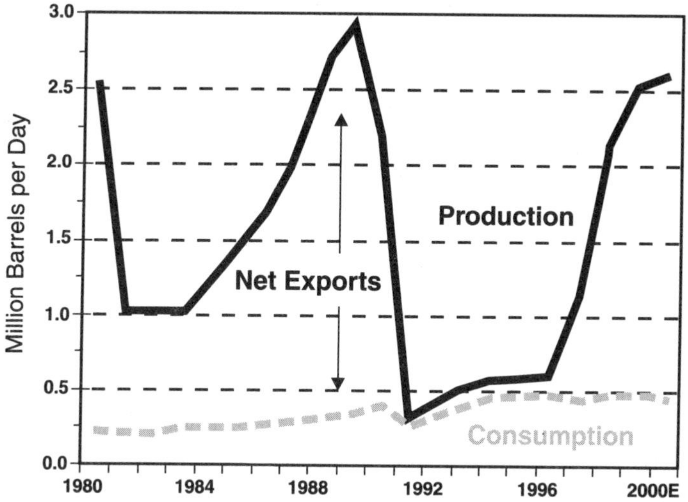

Note: Production includes crude oil, lease condensate, natural gas liquids, ethanol, and refinery gain.

Source: Energy Information Administration, 2001

Prior to the Gulf War, the Rumaila oil field near Basra was producing 1.4 million bbl/d and the Kirkuk oil fields 900,000 bbl/d (EIA, 2001). This level of output fell to approximately 300,000 bbl/d in 1991 and remained there until UNSCR 986 initiated the Oil-for-Food Program (see Figure 9.2). Prior to the war in 2003, Iraq was legally exporting 2.29 million bbl/d of oil through the UN, of which 60 percent of the earnings went directly to food and medicine, 30 percent were directed to the Compensation Commission, and 10 percent funded the UN mission in Iraq (CountryWatch.com, 2001; EIA, 2001). Essentially, this was the extent of Iraqi economic activity, and Iraq could not legally export any commodity without permission from the UN. All oil revenues were funneled through the UN under the terms of UNSCR 986.

The oil that was exported moved via the 600-mile Kirkuk-Ceyhan pipeline connecting the Kirkuk oil fields to Turkey, capable of moving 1.1 million bbl/d. There is a second parallel line designed to carry 500,000 bbl/d of export oil from Basra to Turkey, but this line is currently inoperable (EIA, 2001). In 1975, Iraq finished construction on its 1.4 million bbl/d revisable strategic pipeline. These twin north-south lines allowed Kirkuk to ship oil to Iraq's Persian Gulf ports and Basra to ship to Turkey. (EIA, 2001). This was not the only source of revenue for the Iraqi government under Saddam Hussein. As early as 1997 there were allegations that Iraq had been illegally exporting oil through the 50-year-old Banias pipeline connecting Kirkuk to Syria and Lebanon (EIA, 2001). Syria has stated it would comply with the guidelines established in

UNSCR 986 with regard to any oil it receives from Iraq (CountryWatch, 2001; EIA, 2001). Allegations were also made that truckloads of oil were being shipped over the border by companies controlled by Saddam's family. Iraq's oil infrastructure is shown in Figure 9.3 [C-11].

As of March 2000 Iraq was averaging 2 million bbl/d of legally exported crude oil, most of which was exported through the southern port of Mina al-Bakr, rather than through the Kirkuk-Ceyhan pipeline (EIA, 2001). This made it more difficult for the UN to monitor food-for-oil sales and increased the potential for Iraqi oil smuggling operations via ocean routes. Russian companies, along with Italian, Malaysian, French, and Chinese firms, purchased most of the exported oil for resale to the larger markets in North America, Europe, and Japan. The United States consumed over 25 percent, or 600,000 bbl/d (CountryWatch, 2001; EIA, 2001). Because the purposeful destruction of oil wells by the Hussein regime at the beginning of Coalition attacks in 2003 seem to have been kept to a minimum, Iraq should see resumption of oil exports as soon as the country is stabilized.

AGRICULTURE

Although approximately 12 percent of Iraq's landmass is arable, less than 1 percent of this area is dedicated to permanent cropping (CIA, 2001; CountryWatch, 2001). The lands between the Tigris and Euphrates, once well irrigated and productive, have either been abandoned or have become salinized (FAO, 1997). As Figure 9.4 [C-12] and Table 9.1 show, Iraq once had a fair amount of crop diversity.

The salinization problem, coupled with the urban pull factors of wage labor and service sector employment, has resulted in Iraq becoming dependent upon food imports. As much as 65 percent of Iraq's food supply was imported prior to the Gulf War (FAO 1997). Since the Gulf War and the resulting economic sanctions, Iraq has been unable to purchase adequate food imports to support its population. The UN, in an attempt to mitigate the impending famine, adopted UNSCR 986, the Oil-for-Food Program. As of early 2003, the Iraqi people were dependent upon humanitarian food rations (2,030 kcal and 47 g of plant protein per person per day) purchased from the proceeds of controlled crude oil sales (FAO, 1997). The Iraqi people will remain at risk of famine until a government reclaims the lost agricultural capability and food imports are stabilized. If the Oil-for-Food Program ceases in a post-Hussein Iraq without adequate substitutes, Iraq will be unable to feed its 23.6 million people.

TRANSPORTATION

Iraq has a limited highway system and the majority of its roads emanate out of Baghdad as radial spokes with very few interconnecting beltways. In short, and with exceptions such as Basra and Mosul, all roads lead to Baghdad.

Economic Geography

Table 9.1 Area, Production, and Yield of Cereal Crops

Crop	1996*	1997*	1998*	1999*	2000*	2001*
Barley	1,300,000	778,000	859,000	500,000	226,000	Unknown
Maize	125,000	121,000	133,000	112,000	53,000	60,000
Potatoes	390,000	400,000	420,000	365,000	150,000	150,000
Paddy Rice	270,000	244,000	300,000	180,000	130,000	130,000
Sugar Beets	7,700	7,800	7,850	7,500	7,200	7,200
Soybeans	1,650	1,750	1,780	1,650	1,600	1,650
Sugar Cane	69,000	70,000	71,000	68,000	65,000	65,000
Wheat	1,300,000	1,063,000	1,130,000	1,050,000	555,000	550,000
Total Production	3,463,350	2,685,550	2,922,630	2,284,150	1,187,800	963,850
Percent Growth	N/A	−0.22	0.088	−0.22	−0.48	−0.189

*Metric Tonnes

Prior to war in 2003, Mina al-Bakr was the only operational port that could accommodate supertankers. Iraq's other tanker-capable ports were rendered inoperable during the Gulf War (EIA, 2001). Basra's strategic location at the mouth of the Shatt al Arab section of the Tigris and Euphrates places this city in a position to become Iraq's primary break-in-bulk point. Additionally, the Basra and Umm Qasr dry port facilities were being upgraded prior to hostilities in 2003. Umm Qasr became the first port to reopen after Coalition attacks in March 2003 (Figure 9.5 [C-8]).

CONCLUSION

Iraq has the natural resources to make it one of the richest countries in the region and propel it into the twenty-first century as a modern state. Unfortunately, 20 years of conflict and the associated economic sanctions have wrecked its infrastructure and its ability to capitalize on its oil resources. Because of the actions of the Iraqi government, its people have been relegated to living in one of the poorest states in the region and being dependent upon humanitarian assistance for survival.

10

Population and Urban Geography

Dennis D. Cowher
Brandon K. Herl

Key Points

- Iraq's population is not uniformly distributed; it is clustered in the cities of Baghdad, Mosul, and Basra
- Iraq's population is growing fast, creating a large population under age 14
- Demographic statistics reveal the poor quality of health care in Iraq

Harsh climates and limited water availability directly affect Iraq's population patterns. In short, Iraq's 23.6 million people are not distributed uniformly across the country. As Figure 10.1 [C-13] indicates, the southwestern half of Iraq west of the Euphrates River is very sparsely populated, home to only 0–65 persons per square mile. The majority of Iraq's population lives in the floodplains of the Tigris and Euphrates Rivers, concentrated in and around the capital of Baghdad. The provinces of Babil and Karbala, located south of the capital, are also densely populated relative to the rest of the country. These areas contain between 194 and 453 persons per square mile.

Overall, the majority of the Iraqi people live in cities, with 76 percent of the population considered urban and only 24 percent rural (Microsoft Encarta, 2001). This makes Iraq about as urbanized as the United States.

Iraq has three cities with populations greater than one million people. The largest, with as many as 5.6 million, is Baghdad, the capital; the remaining two are Mosul, with 1.7 million, and Basra with 1.3 million people. Arbil, Kirkuk, and Sulaimaniya are the next three largest settlements. The top ten cities are listed in Figure 10.1 [C-13].

BAGHDAD

One of every five to six Iraqis lives in the capital, Baghdad. Iraq's largest city, located astride the Tigris River, Baghdad has a population of over 5.5 million people. It is also the center of air, road, and railroad transportation in the country. As the leading manufacturing city, it has numerous industrial facilities supporting the petroleum industry as well as food-processing plants, textile factories, and other industries. Like many riparian cities, Baghdad throughout history accepted both the blessings of the river and its periodic wrath during floods. Today, a dam north of the city has greatly reduced the hazard of flooding. The greater Baghdad area is presented in Figure 10.2 [C-14].

Like many settlements in the region, Baghdad is a city with an important history. In 762 the location was chosen for the capital of the Abbasid Caliphate, a powerful Islamic state that ruled for centuries. Located behind three concentric walls, the Round City, over 2,500 m (8,125 ft.) in diameter, served as home for the caliph, his family, staff, and servants. Naturally, the area around the walled compound grew as well, and soon the Baghdad area was a large metropolis.

The eighth and ninth centuries of Abbasid rule are often referred to as the golden age of Islamic culture, and Baghdad was at the heart of this flourishing scene. Many of the stories of *The Thousand and One Nights (Arabian Nights)* take place in Baghdad during this era. In addition, trade flourished, and goods from as far away as India, China, and Africa could be obtained in Baghdad.

As a city, Baghdad declined between the ninth and twelfth centuries before brutal attacks by Mongol invaders in 1258. Led by Hulagu Khan (Hülegü), a grandson of Genghis Khan, the invaders razed the city and killed over a hundred thousand people. In addition, destruction of irrigation systems reduced Baghdad's ability to regain its past glory. The city was sacked again in 1401 by Timur (Tamerlane), the powerful ruler of a Central Asian Mongol subgroup based in what is now Uzbekistan. Again, tens of thousands were killed and the city laid to waste. By the early sixteenth century the city was under control of the Ottomans after changing hands regularly. By the nineteenth century, the British had established an active diplomatic presence and Ottoman modernizations brought increased foreign trade, and thus prosperity, to the city. It was thus a natural choice for a modern capital after the state of Iraq was created by League of Nations mandate after World War I. Bombing during the 1991 Gulf War destroyed much of the city's infrastructure and rebuilding efforts were constrained by UN economic sanctions and misuse of funds by the Iraqi regime. The bombing in spring 2003 will again necessitate years of rebuilding.

The human landscape of Baghdad reflects both the old and new. In some districts, older mud houses and narrow streets remind the visitor of the city's past, as do mosques and other buildings from the Abbasid and Ottoman periods. Elsewhere, the city is quite modern and features large monuments constructed by the Baath Party and Saddam Hussein.

As mentioned above, Baghdad's population exceeds 5.5 million and mirrors the diversity of the country. Much of the population is Arab and Muslim, from both the Shia and Sunni communities, but other Iraqi minorities, such as Kurds and some Christian groups, also have a presence. As an important city in the region, there are also large Arab populations from countries such as Egypt and Jordan. Although neighborhoods are often delineated by ethnic affiliation, socioeconomic stratification is also a factor in determining where people live.

MOSUL

Mosul is Iraq's second-largest city with a population estimated at 1.7 million (CIA, 2002). The Mosul of bygone eras was a major stopping point along the caravan routes between the Mediterranean Sea, Persia (modern-day Iran), and India. Specifically, it was an important crossing point of the Tigris River. As a testament to its former livelihood as a trade city, older ruins reveal walls similar to those of other cities along the caravan routes to Asia (Arabnet, 2002).

Today, Mosul largely serves the needs of agricultural production in the surrounding districts. Its primary revenues come from cotton production and related textile production (CountryWatch, 2002). This region of Iraq is also home to the most fertile nonirrigated arable lands in the country because of higher annual rainfall totals. The Hussein government's indifferent, and occasionally hostile, relationship with the ethnic Kurds of this region made the area in and around Mosul a potential flashpoint for civil unrest (Encyclopedia Britannica Online, 2001; CIA, 2001; CountryWatch, 2002).

Although Mosul is not technically at the headwaters of the Tigris River, the lands around Mosul contribute a great deal to the river's flow. Because of the relative abundance of water, and thus agricultural production, Mosul and its hinterland may hold an important key to Iraq's long-term survival. The greater Mosul area is depicted in Figure 10.3 [C-15].

BASRA

Basra (Figure 10.4 [C-16]) is Iraq's third-largest city, with over 1.3 million people, and the country's main port. Its location near the confluence of the Tigris and Euphrates Rivers, great marshes, and the Shatt al Arab waterway, give the city great strategic importance both militarily and commercially (CountryWatch, 2002). Indeed, its strategic location was important in its founding in A.D. 637 as both a military base and a departure point for Far East trade. Many of the tales of Sinbad have this legendary sailor frequenting the port during his many journeys (Arabnet, 2002). Outside of legends, Basra has been one of the most fought-over cities in the region over the last 500 years. The fierce fighting in spring 2003 once again emphasized the desire of governments to hold this vital city.

Basra controls the entry and exit to the lands upriver of the Tigris and Euphrates. Because the great rivers are not navigable by large ocean-going vessels upstream from Basra, the city serves as the country's major port. In addition, in the twentieth century Basra became an important oil-refining and export center. As such, control of Basra amounts to control of Iraq's access to the sea as well as access to major oil and natural gas fields. No matter who rules Iraq, they will likely continue to view the Basra region as a life-and-death point of contention in any conflict.

RURAL POPULATIONS

Over 5 million Iraqis are considered rural. Many people in the rural areas of the country still live in isolated communities and lead a nomadic or seminomadic existence, their livelihoods based on herding camels, horses, and sheep. Population increases, however, reduce access to adequate grazing lands, making rural life even more difficult for many. UN sanctions have also impacted these communities.

POPULATION DENSITY

One way to explore population is in terms of density, a numerical measure of the relationship between the number of people and some other unit of interest expressed as a ratio. For example, crude density (sometimes referred to as arithmetic density) is probably the most common measurement of population density. Crude density is the total number of people divided by the total land area.

According to the Population Reference Bureau (2003), Iraq had an estimated population of 23,600,000 in 2002. These people inhabit a state with a land area of 169,236 sq mi, slightly larger than the state of California. Iraq's population density is thus equal to about 134 persons per sq mi. For the sake of comparison, California has a population density of 212 persons per sq mi. These data do not reflect an important concern for Iraq, the amount of arable land in the country. According to the Central Intelligence Agency, only 12 percent of the land in Iraq is considered arable (CIA *World Factbook*, 2002). Therefore, when we calculate the "physiologic density," which takes into account the arable land, Iraq has a population density of 1,204 persons per sq mi of arable land. This is a much more useful measure for comparison. Iraq's measure of 1,204 is roughly three times as many people per sq mi of arable land than the United States. This physiologic density suggests a condition in Iraq in which there is tremendous stress on the country's farmland to feed the growing population. Iraq's farmland is declining in productivity due to soil salinization, which is caused by insufficient drainage and by saturation irrigation practices. In addition to the 12 percent of the land that is arable, 8 percent of Iraq is irrigated farmland. Government-sponsored water control projects, however, have destroyed wetland habitats in the eastern region of the country by diverting or drying up tributary streams that formerly irrigated wetland areas. Moreover, the challenge to feed the growing population will increase the amount of land that must be irrigated. This will increase Iraq's need for water—a scarce commodity in this region of the world. This need for irrigation could possibly result in future water conflicts with neighboring countries, especially Turkey.

POPULATION STRUCTURE

In addition to exploring patterns of distribution and density, population geographers also examine population in terms of its composition, that is, in terms of its subgroups. Understanding population composition enables analysts to gather important information about population dynamics. For example, an analysis of the composition of a population in terms of the total number of males and females, proportions of old people and children, and number and percentage of people active in the workforce provides valuable insights into the ways in which the population behaves.

The most common way for demographers to graphically represent the composition of the population is to construct an age-sex pyramid, which is a representation of the population based on its composition according to age and sex. Age categories are ordered sequentially from the youngest, at the bottom of the pyramid, to the oldest, at the top. By moving up or down the pyramid, one can compare the opposing horizontal bars to assess differences in frequencies for each age group. A cohort is a group of individuals who share a common temporal demographic experience. A cohort is not necessarily based on age, however, and may be defined according to criteria such as time of marriage or time of graduation.

Age-sex pyramids can reveal the important demographic implications of war or other significant events. Moreover, age-sex pyramids can provide information necessary to assess the potential impacts that growing or declining populations might have. The age-sex pyramid for Iraq is shown in Figure 10.5.

Figure 10.5 reveals two interesting points that must be considered in more detail. The first and perhaps most important point to make about Iraq's population composition is that nearly half of Iraq's population is below age 15. This is a very high percentage compared to the world average of 30 percent. Iraq's percentage, however, is comparable to other fast-growing Arab states in the region, such as Jordan, Saudi Arabia, and Syria, which all have values exceeding 40 percent.

What are the implications of this age composition? The considerable narrowing of the pyramid toward the top indicates that the population of Iraq has been growing very rapidly in recent years. The shape of Iraq's profile is typical of countries with high birthrates and relatively low death rates.

A high rate of population growth leads to a situation in which the ratio of workers (people of working age) to dependents (people either too young or too old to work) is much lower than if a population is growing slowly. This means

Figure 10.5 Population Pyramid of Iraq, (2000)

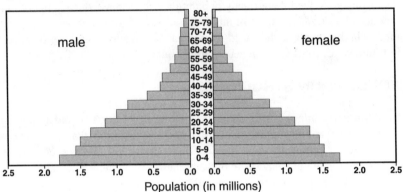

Source: U.S. Census Bureau, International Database, 2002

that in a rapidly growing society, each worker will have to produce more goods just to maintain the same standard of living for each person as in a more slow-growing society. This may seem like an obvious point—the father of six children will have to be more productive (or earn more money) than the father of three just to keep his family living at the same level as the smaller family—but it goes deeper than that. A state depends at least partially on savings from within its population to generate investment capital with which to expand the economy, regardless of the kind of political system that exists. With a very young age structure, money gets siphoned into taking care of more people (buying more food and providing education services for example) rather than into savings per se.

A second concern for countries with fast-growing populations is the number of prospective entrants into the labor force. If economic development is to occur, the number of new jobs must at least keep pace with the number of people looking for them. The expansion of jobs is, of course, related to economic growth, which in turn relies on investment and may be harder to generate with a young age structure. There are several economic and political reasons why Iraq's economy is not growing fast enough to keep up with its population growth. This chapter will not detail these, but in short, two recent wars have severely reduced economic activity. In addition, UN-imposed economic sanctions since the end of the Gulf War proved to be an obstacle to the growth that is essential for Iraq's expanding population.

BIRTH RATES AND DEATH RATES

The Crude Birth Rate (CBR) is the total number of live births in a year for every thousand people in the population. The CBR for Iraq in the year 2002 was 35 (Population Reference Bureau, 2003). This CBR was 30 percent above the regional average. Only Saudi Arabia has a higher birthrate. To give this number some perspective, consider that it is over twice the number of the United States, which is about 14. Although the level of economic development is a very important factor shaping the CBR, other, often equally important influences also affect CBR. The case of Iraq is similar to other Arab-Muslim states in Southwest Asia. Saudi Arabia, Jordan, and Syria, all bordering Iraq, have similarly high birthrates. These data suggest that high birthrates are not simply an Iraqi phenomena. Rather, high birthrates in the Arab-Muslim world reflect religious, social, and other cultural factors. Low levels of female education have been repeatedly linked to higher birthrates. For example, in Iraq only 46 percent of women in the country are literate. This compares with a male literacy rate of 66 percent (Population Reference Bureau, 2003). Diet and health, war, and other political unrest are also mentioned by scholars as reasons why birthrates are higher than average in the Arab-Muslim world. But it should be made clear that Iraq has a high CBR even by Arab-Muslim standards.

The CBR is only one of the indicators of fertility. Another indicator used by population experts is the Total Fertility Rate (TFR), which measures the average number of children a woman will have throughout her childbearing years, generally considered to be ages 15 to 49. Whereas the CBR indicates the number of births in a given year, the TFR is a more predictive measure that attempts to portray what birthrates will be among a particular cohort of women over time. A population with a TFR of slightly higher than two has achieved replacement-level fertility. This means that birthrates and death rates are approximately balanced and there is stability in the population. The TFR for Iraq is 5.4, which is well above the regional average of 3.9 (Population Reference Bureau, 2003). In the United States and many other developed countries, the TFR is 2.1 or below. It should be noted that the TFR in Iraq has fallen from 6.6 in the late 1970s to its current rate of 5.4. But this current TFR remains well above the world's average rate of 2.8. Iraq's TFR indicates that the population will continue to grow rapidly in the future. The U.S. Census (2002) predicts that Iraq's future TFR will continue to decline, reaching a more stable figure of 2.7 by the year 2025. Unfortunately for Iraq, even with this drop in fertility, the current population could grow to as high as 60 million by the middle of the century (Population Reference Bureau, 2003).

Another measure of population growth is the "doubling time" of the population. The doubling time, as the name suggests, is a measure of how long it will take the population of an area to grow to twice its current size. To compute a country's doubling time, we simply divide the number 70 by the rate of natural increase. In the case of Iraq, the rate of natural increase is 2.5 percent, over four times that of the United States. To calculate Iraq's doubling time, we divide 70 by 2.5, and we get a period of 28 years. It is troubling, given Iraq's current level of economic development and environmental concerns, that the country's population may double to nearly 50 million in the next two to three decades.

Countering birthrates and also shaping overall population numbers and composition is the Crude Death Rate (CDR), the total number of deaths in one year for every thousand people in the population. Mortality rates often reflect levels of economic development, with higher rates indicating less developed countries. Iraq's CDR is 10, which is above the regional average of 7 per 1000.

Death rates can be measured for both sex and age cohorts and one of the most common measures is the Infant Mortality Rate (IMR). This figure is the annual number of deaths of infants less than one year of age compared to the total number of live births for that same year. The figure is usually expressed as number of deaths during the first year of life per 1,000 live births. The IMR has been used by researchers as an important indicator both of a country's health care system and the general population's access to health care. Iraq's IMR was 103 deaths per 1,000 live births in 2002, which is much higher than the regional average of 45 infant deaths per thousand live births (Population Reference Bureau, 2003).

Related to infant mortality and the CDR is life expectancy, the average number of years a newborn can expect to live. Infants born in Iraq in the year 2002 can expect to live an average of 58 years, while infants born in the same year in neighboring countries can expect to live an average of ten years longer, indicating poorer than average health care (Population Reference Bureau, 2003). Unfortunately, these statistics are only given at the national level for Iraq. It would be interesting to note differences among Iraq's different minority groups and regional differences as well. Provincial data on these statistics are not available. It is important to note a finding of the U.S. Committee for Refugees, that 800,000 Iraqi children under age five were chronically malnourished and that 10 percent of children under age five in Baghdad, Karbala, and Diyala indicated "wasting" (low weight for height). On the other hand, the three Kurdish-controlled northern provinces appeared to be enjoying relative prosperity, both as result of receiving a UN-mandated 13 percent of all Oil-for-Food revenues as well as "taxes" the Kurds impose on the lucrative smuggling operations across the Turkish and Iranian borders (USCR Country Reports, 2002).

MOBILITY AND MIGRATION

In addition to the population dynamics of death and reproduction, the movement of people from place to place is a critical aspect of examining population geography. Mobility is the ability to move from one place to another, either permanently or temporarily. There remains a nomadic segment of the Iraqi population. Even more important for population geographers is the millions of marginalized Kurds and Shiite Muslims who have been forced to move in the past decade because of political, economic, and environmental factors. A good example is the sizable population of Shiite Muslims in the south who have been displaced because government water control projects have drained most of the inhabited marsh areas east of Nasiriyah (CIA *World Factbook*, 2002).

The second way to describe population movement is in terms of migration, which is a permanent move to a new location. Migrants permanently change their place of residence—where they sleep, store their possessions, and receive legal documents. Migration has two forms, emigration and immigration. Emigration is migration out from a location; immigration is migration into a location. A decision to migrate stems from a perception that somewhere else is a more desirable place to live. People may hold very negative perceptions of their current place of residence or very positive perceptions about the attractiveness of somewhere else. Negative perceptions about their place of residence that induce people to move away are push factors, whereas pull factors attract people to a particular new location.

Migration from Iraq is basically resulting from three push factors: political, economic, and environmental. The Kurds and Shiite Muslims make up the majority of the population of Iraq, yet they had little say in the government of Saddam Hussein, who repeatedly repressed these groups when they attempted to gain more autonomy. The environmental degradation ongoing in Iraq is another reason many people have decided to emigrate. Refugees are people forced to migrate from a particular country for political reasons. The UN defines political refugees as people who have fled their home country and cannot return for fear of persecution because of their race, religion, nationality, and membership in a social group, or political opinion (Rubenstein, 1996). According to the U.S. Committee for Refugees (USCR), there were more than 127,700 refugees and about 700,000 internally displaced persons in Iraq in the year 2000 (USCR, 2002). Foreign refugees in Iraq include about 23,900 people from Iran and 12,600 from Turkey, mostly Kurds. The total also included about 90,000 Palestinians and about 1,200 refugees of other nationalities, including Eritreans (600), Somalis (300), Sudanese (200), and Syrians (100).

The estimated 600,000 internally displaced persons in the three northern provinces of Dohuk, Erbil, and Sulaimaniya included not only long-term internally displaced persons and people displaced by Kurdish factional fighting, but also as many as 100,000 people, mostly Kurds, Assyrians, and Turkomans, more recently expelled from the central-government controlled Kirkuk and surrounding districts in the oil-rich region bordering the Kurdish-controlled north. At least another 100,000 persons were internally displaced elsewhere in Iraq, mostly in the southeastern marshlands.

Between 1 and 2 million Iraqis with a well-founded fear of persecution were estimated to be living outside Iraq, although only about 550,000 had any formal recognition as refugees or asylum seekers in 2000. About 510,000 Iraqi refugees were living in Iran and about 5,200 refugees remained in the Rafha camp in Saudi Arabia. In 2000, some 34,000 Iraqis applied for asylum in Europe.

CONCLUSION

Iraq's demographic statistics paint a bleak picture of the future prospects for the country. Iraq's economy does not currently have the capacity to grow fast enough to provide opportunities to its fast-growing population. If natural growth rates are not lowered by aggressive policies, Iraq will likely require international aid to avoid famine and human misery on a massive scale. The segments of population that are most vulnerable in Iraq are marginalized groups, including 4 million Kurds and 13 million Shiite Muslims. The Iraqi government has certainly been part of the problem, and it is often reported that Saddam Hussein spent money on palaces and military programs that could have been used to provide infrastructure improvements, health care, quality food, and new irrigation schemes.

11

Medical Geography

Patrick E. Mangin

Key Points

- Infectious diseases are endemic
- Public health systems are in complete disarray, increasing disease and mortality rates
- Food and water in Iraq is often unsafe
- Poor nutrition is a major cause of health problems
- Heat and cold also pose a human health risk in the area

A valuable resource for understanding the dynamics of a country is its medical geography. Medical geography by definition is "the application of geographical perspectives and methods to the study of health, disease and health care" (Johnson, 1996). Medical geography incorporates two broad areas of study. The first concerns the spatial ecology of disease and geographical aspects of the health of populations. The second emphasizes the geographical organization of health care. Medical geography retains associations with other disciplines outside geography concerned with health-related problems, reflecting the complexity of these problems and the need to examine them from a multidisciplinary perspective. This chapter will analyze the overall health of the Iraqi people based on the distribution of disease and nutrition.

THE TRIANGLE OF HUMAN ECOLOGY

A useful framework for analyzing the impact of health-related issues is the triangle of human ecology (Figure 11.1). Three vertices form a triangle of population, behavior, and habitat, which together affect a region's health (Palka, 2001). As Palka writes,

> Habitat is that part of the environment within which people live. It includes houses, workplaces, agricultural fields, recreation areas, and transportation systems. Population considers humans as the potential hosts of various diseases. Factors affecting and yet characterizing the population include nutritional status, genetic resistance, immunological status, age

73

Figure 11.1 The Triangle of Human Ecology

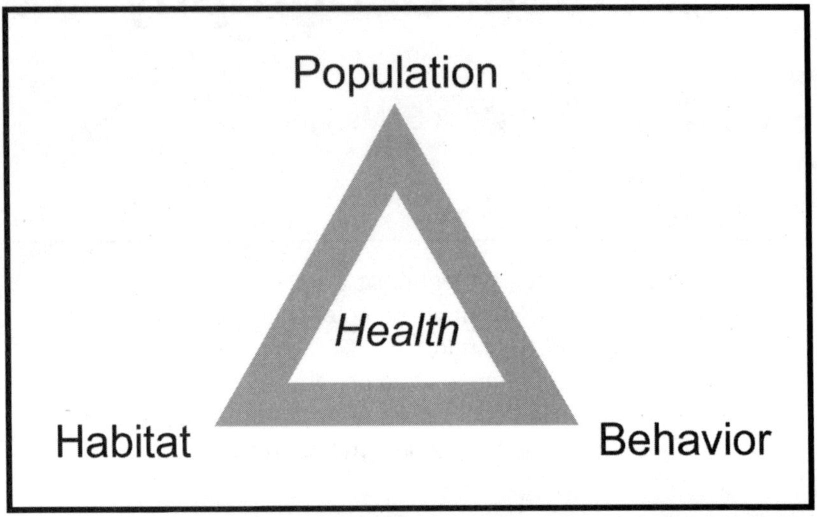

Population

Health

Habitat   Behavior

Source: Palka, 2001

structure, and psychological and social concerns. Behavior includes the observable aspects of the population and springs from cultural norms. It also impacts on those who come into contact with disease hazards and whether or not the population elects other alternatives. (Palka, 2001)

Health is a state of complete physical, mental, and social well-being, and not merely the absence of disease. Health is a continuing property that can be measured by an individual's ability to rally from a wide range and considerable amplitude of insults. (Johnson, 1996)

Chemical insults include pollen, asbestos, various pollutants, smoke, waste, or even food (Johnson, 1996). Infectious insults include virus, bacteria, fungi, and protozoa. Infectious insults cause debilitating endemic and epidemic diseases. Physical insults could refer to air quality, temperature, humidity, light, sound, atmospheric pressure, and trauma. Physical insults unique to Iraq include the stress of extreme annual and diurnal temperatures, frequent dust storms, and an arid climate that could contribute to dehydration.

THE IMPACT OF DEPLETED URANIUM MUNITIONS FROM THE 1991 GULF WAR

The 1991 Gulf War had both immediate and residual negative impacts on the people of Iraq from a health perspective. The obvious immediate effect of the war was the decrease in Iraqi population due to combat deaths. In addition to having a direct impact in the hostilities, the residual effects of Allied munitions on the health levels of the Iraqi people are less obvious, but equally serious, to

Medical Geography

Figure 11.2 Primary Areas of Depleted Uranium Expenditure During the Gulf War

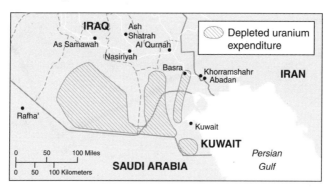

Source: Department of Defense, 1998; CIA basemap

the Iraqis who survived the war and to subsequent generations. Unexploded ordnance pose a risk, and additionally, the presence of depleted uranium (DU) rounds could represent a risk to the overall health of the Iraqis, especially children. Depleted uranium rounds fired from Allied tanks and infantry vehicles are scattered about southern Iraq (Figure 11.2).

Data concerning depleted uranium in Iraqi health should be approached cautiously. While there have undoubtedly been declines in Iraqi health, there is little evidence in any peer-reviewed sources that suggests a cause-and-effect connection between DU and cancer in southern Iraq. Based on available reports, DU is certainly radioactive and if ingested, inhaled, or handled for long periods (weeks), there could be long-term health effects.

THE HEALTH AND MEDICAL IMPACT OF UN SANCTIONS

Another Gulf War-related factor contributing to the decline in health is UN Resolution 661 of 1990, which effectively cut off all shipments of commodities and products to Iraq (United Nations, 1990). Medical supplies (medicine, vaccinations, and medical item repair parts for clinics and hospitals) and food could no longer be imported into Iraq. This was done to reduce Iraq's combat effectiveness. The immediate impacts of the sanctions were readily seen in the steady decline in the health of the Iraqi population, especially children. According to a survey by the United Nations Children's Fund (UNICEF), though the death rate of Iraqi children five years and younger steadily declined during the 1980s, it more than doubled to 120 or 130 persons per thousand during the 1990s after the Gulf War (UNICEF 1999). Figure 11.3 suggests that Iraq showed a substantial reduction in the under-five mortality rate during the

Figure 11.3 Infant and Child Mortality Rates, 1960–1998

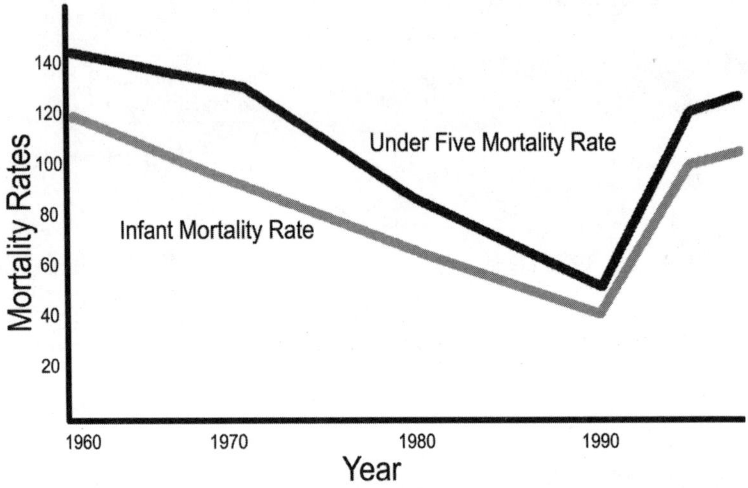

Source: Based on data from UNICEF, 1999

1980s. Had this trend continued through the 1990s, there would have been half a million fewer deaths of children under age five in the country as a whole during the eight-year period 1991–1998 (UNICEF 1999). Therefore, there seems to be a clear correlation between increased child mortality and the lack of food and medical supplies stemming from the UN sanctions. It should be noted, however, that resources that could have been used for health infrastructure were often diverted by the regime to further their own agenda. Many presidential palaces were constructed during the same period as health was declining.

In the early 1990s, Iraq's food supply situation became a critical problem because of food sanctions. Because it is an arid country with limited agricultural opportunities and a growing population, Iraq is a net importer of foodstuffs, such as cereals (Figure 11.4). Prior to the war, Iraq had to import a significant amount of grains, vegetables and fruit to support itself. After the cease-fire, Iraq could not import these goods, so nutrition suffered. In light of this, in 1992, the UN Security Council adopted Resolution 706, "offering an opportunity for Iraqi oil to be sold and the revenue used to purchase essential humanitarian supplies" (UNOIP, "Implementation" 2002). Unfortunately, the Iraqi government refused the terms of the resolution and food sanctions continued mainly because Saddam Hussein's regime was not cooperative with UN inspectors, who were not allowed to inspect many places thought to be manufacturing weapons of mass destruction (WMD). The WMD inspections were a part of the 1991 cease-fire agreement between the Allies and Iraq.

Figure 11.4 Net Cereal Imports as a Percentage of Total Cereal Production

Source: Based on data from the World Resources Institute, 2000

AGRICULTURAL PRODUCTION

Owing to its arid climate, Iraq is agriculturally constrained to riparian (near rivers) areas and therefore irrigates otherwise nonarable land. Irrigated cropland makes up 60 percent of Iraq's available cropland. Irrigation, though, does not allow Iraq to support its food needs.

The amount of land Iraq can irrigate has significantly declined in the past decade owing to Turkey's Anatolia Project. The Grand Anatolia Project is a multiple Turkish dam project designed to improve Turkey's hydroelectric and irrigation capability. The Turkish government has built 22 dams on the Tigris and Euphrates Rivers, which reduce the volume of water flowing into Iraq. The upstream dams reduce the volume flowing into Iraq in three ways. First, the increased surface area of the reservoirs created by the dams allows for more evaporation. Second, the water becomes still, and a higher hydraulic head is created, forcing more groundwater seepage. Third, the water in the reservoirs is pumped throughout Turkey for agricultural irrigation and industrial uses.

Reduced water flow has negatively impacted Iraq's agricultural production. The marshes of southern Iraq, an important agricultural area, are drying up in part from less water entering them. The amount of water available to irrigate other areas is reduced. This has meant less food for Iraq in the post-Gulf War period. To make matters even worse, Iraq suffered a three-year long drought beginning in 1999.

Medical Geography

Table 11.1 Nutrition Indicators, 1989 and 1999

Indicator	1989	1999
Calories/day	3,089 calories/day	2,100–2,200 calories/day
Birth Weight < 2.5 kg	102 g/day	46–47 g/day
Kwashiorkor	4.5%	24%
Marasmus	41/month	2,091/month
Underweight	1–2%	21%
Proportion of families consuming animal protein	100%	46%

Source: Iraq Ministry of Health, 2001

The impact of less water has meant less food and this leads to undernutrition and malnutrition. Combined, undernutrition and malnutrition contribute to poor health (increases in disease occurrence) and increased death rates (especially in children). Undernutrition means that the Iraqis do not get the necessary caloric intake to maintain their health, and malnutrition means that the Iraqis do not get well-balanced sources of nutrition necessary to stay healthy and fight off disease. Iraq Ministry of Health statistics indicate that the caloric intake of the Iraqi people is down one-third from pre–Gulf War levels. An individual's ability to rally from a wide range and considerable amplitude of insults is directly related to nutrition, and Table 11.1 reflects an overall increase in the Iraqis succumbing to various diseases owing to malnutrition and undernutrition. The table also shows that a significant fall in newborn birth weights as well as increased infant cases of kwashiorkor (protein deficiency) and marasmus (malnutrition). The infant mortality figures indicate that in 1989, there were 7,110 infant deaths under the age of five; in 1999, there were 73,572 (Iraq Ministry of Health, 2001).

Eventually, Iraq agreed to the Oil-for-Food Program. At the end of 1996, "the United Nations and the Government of Iraq agreed on the details of implementing resolution 986 (1995), which permitted Iraq to sell up to two billion dollars worth of oil in a 180-day period" in order that Iraq could buy food (UNOIP, "Oil-for-Food" 2001). The ceiling on oil sales was eased during 1998 and finally lifted in 1999, enabling the program to move from a focus on food and medicine to repairing essential infrastructure, including the oil industry. Declines in public health infrastructure, such as liquid waste disposal and water treatment can directly relate to the type of diseases seen in Table 11.2.

In the petrochemical sector, oil was first exported on 15 December 1996 and the first contracts financed by the sale of oil approved in January 1997. The first shipments of food arrived in Iraq in March 1997 and the first medicines arrived in May 1997. The Security Council continued the program in 180-day periods called "phases."

Table 11.2 Disease Incidence, 1989 and 1999

Disease	1989 (# of cases)	1999 (# of Cases)
Typhoid	1,812	23,392
Cholera	0	2,398
Poliomyelitis	10	75
Diptheria	96	142
Measles	5,715	9,920
Pertussis	368	466
Tuberculosis	14,312	29,897

Source: Iraq Ministry of Health, 2001

Since the first food arrived in March 1997, foodstuffs worth over $8 billion have been delivered to Iraq. Although it is difficult to assess the impact of the program, the average daily food ration gradually increased from around 1,275 kilocalories per person per day in 1996 (before the program) to about 2,229 kilocalories in October 2001 (UNOIP, 2001).

MEDICAL SUPPLIES

Sanctions also contributed to declining availability of health care supplies. There are shortages of the most basic medical supplies. Everything from bandages, sheets, rubber gloves, and syringes are carefully managed or are even dangerously reused. Sanitation equipment is faltering owing to a decline in repair parts. Vaccination supplies are low, contributing to the increased cases of diseases shown in Table 11.2. A lack of vaccinations contributes to people succumbing to common communicable diseases that they would normally ward off. "Common communicable diseases preventable by vaccination, which are provided [to] nearly all children in developed countries and were standard in Iraq before 1990, have increased by multiples" (Clark, 2000). While occurrence rates for these diseases fluctuate, those of malnutrition-related diseases and death rates in general do not; because of the cyclical nature of their communication, they have increased at a regular rate, and afflicted an additional hundreds of thousands of children. For example, "Increases in 1998 over 1989 were as follows: whooping cough, 3.4 times; measles, 4.5 times (25,818 cases); mumps, 3.7 times (35,881 cases)" (Clark, 2000).

Not only are medicines and vaccination in short supply, but less obvious items are also lacking and contribute to poorer health. One writer described the hospitals as lacking in the most basic of cleaning and medical supplies and struggling with electricity problems. "They [the hospitals] are infested with flies and smell of urine and feces. The electricity is turned off periodically each day to conserve. Due to a lack of supplies and medicines, patients succumb everyday to illnesses that normally are easily treated" (Johnson, 1999). Clinics and hospitals that cannot disinfect instruments and wards will likely see increased levels of disease.

Since 1997, under the Oil-for-Food Program, health supplies worth about $1.5 billion have been bought, but health trends still show significant declines.

CULTURAL INFLUENCES ON THE HEALTH OF IRAQIS

Iraq's population growth continues to detract from its health. The population growth rate is estimated at 2.5 percent (Population Reference Bureau, 2003). A growth rate of 2.5 percent means that Iraq's population may double in less than 30 years. Between 1995 and 2000, the average total fertility rate was 5.3. This means that the average woman between the ages of 15 and 49 has over five children. A growing population under UN sanctions means more mouths to feed and a declining amount of food available. The burgeoning population is not unrelated to cultural influences such as Islam. Birth control is a good example, as the following quotation highlights:

> Birth control is virtually non-existent in Iraq, as limiting births or interfering with conception in any way is against the laws of Islam. Life is considered a gift from God. Likewise, abortion in any form would be out of the question. However, the Iraqi people are beginning to understand that their population growth is unsustainable and have begun to use; [modern contraception]. Even among husbands, support for birth control is growing. There is an acceptance, or at least a rationalization, that limiting births is a means for adaptation and economic sufficiency. (Kemp, 2000)

Iraqi adherence to Islamic dietary laws is an area where Iraqi cultural beliefs contribute to better health. *Halal* foods (legal in the religious sense) contribute to Iraqi health from a nutrition standpoint. Briefly, any meat consumed by a Muslim must come from an animal slaughtered by another Muslim in a prescribed way, or it is considered impure, *haram* (illegal in the religious sense). This means that their meat is usually well prepared. Furthermore, pork and alcohol are especially haram and are not consumed by most Iraqi Muslims. The immediate impact of no pork or alcohol is the generally low heart and lung disease in Iraq. This is quite opposite of the United States, for example, in which these diseases are leading causes of death.

WATER SUPPLY AND HEALTH

The living and sanitary conditions in Iraq can be summarized from an excerpt from the Defense Intelligence Agency's "Medical Environmental Disease Intelligence and Countermeasures" CD-ROM, published before Coalition attacks in 2003:

Medical Geography

Living and sanitary conditions are poor and continue to deteriorate. Expanding slum areas place additional demands on an already overburdened urban infrastructure. Shelters constructed by migrants on vacant lots in slum areas house an average of six persons in a single room. Urban sewage floods homes and streets because of inoperative electric pumping stations. Untreated waste water discharges into surface water sources.

Iraq's draining of Shiite-inhabited southern marshes is expected to cause the die-off of reed beds upon which the inhabitants depend for house-building materials, cattle fodder, and fuel. As the government of Iraq continues to drain marshes, it is expected that more refugees will continue to flow to Iran.

Raw sewage contamination of water sources is the most significant pollution problem. Disrupted irrigation systems are resulting in desertification in many areas of the country. Movement of heavy weapons and troops has caused extensive soil damage, particularly in the fragile desert soil near Saudi Arabia. (DIA-AFMIC, 2001)

About one-half of Iraq's population obtains water directly from surface sources, such as rivers, reservoirs, irrigation canals, drainage ditches, and open wells. The remaining one-half depends on municipally supplied and bottled water (DIA-AFMIC, 2001). Water and sewer facilities sustained severe damage during the course of military operations in the 1991 Gulf War. Air and ground forces destroyed or damaged many water treatment plants, pumping stations, laboratories, and other related equipment, particularly in the southern provinces. Therefore, after the war, a primary mission of Iraq's government was to rehabilitate the water supply and sewage systems. "Prior to the Gulf War, municipal water systems provided treated water to 95 percent of the urban population, but only to 40 percent of rural populations" (DIA-AFMIC, 2001). Due to the sanctions, the construction of 18 water treatment projects was halted because of a lack of necessary building materials. These projects were under construction when the aggression started and were 45–95 percent from being complete. The total annual capacity of these projects was 980 million cu m/year representing 48 percent of the present total water supply capacities (DIA-AFMIC, 2001)

There is good news, however, in that Iraq has rehabilitated previously existing water treatment plants to about 50–60 percent of their prewar capacity. Keeping them operational is difficult, owing to a lack of repair parts; for example, instead of replacing leaking pipes, duct tape is often used as a permanent solution (Iraqi Ministry of Health, 2001). The sanctions impact rural inhabitants the most. Large numbers of people in the rural areas are without access to safe drinking water or receive it in inadequate quantities. The situation is further aggravated by a severely limited availability of purification chemicals and poor maintenance of purifying equipment. "Chemical plants producing alum, chlorine, and sulphate needed to treat water were damaged during

Figure 11.5 Water Supply and Sanitation Programs, 1999–2001

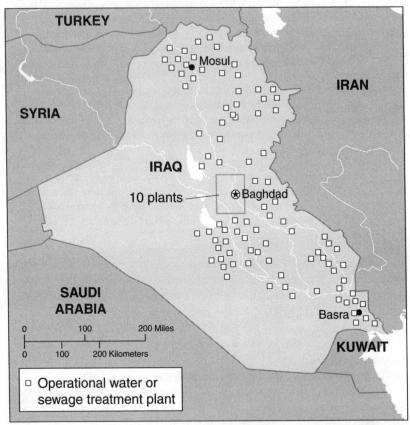

Source: Based on data from Relief Web, 2001

the war," and their production capacity remains limited by the sanctions" (DIA-AFMIC, 2001). As a result of the Oil-for-Food Program, the Iraqi government began to reinvest in water sanitation programs (Figure 11.5). Until these projects are completed in a post-Hussein Iraq, most rural areas will continue to suffer from contaminated water.

MAJOR WATERBORNE DISEASES

"During the World Health Organization's decade of water development in the 1980s, water-related diseases were classified into three development-related categories: waterborne diseases, which were ingested; water-washed (or un-washed) diseases, which were preventable by hand/hair/clothes/floor washing and other hygiene; and water-based diseases, which were vectored diseases re-

Medical Geography

Table 11.3 Major Water-Related Diseases

Waterborne	Water-unwashed	Water-based
Diarrhea	Intestinal worms	Malaria
Typhoid	Amebic dysentery	Filariasis
Cholera	Colds	West Nile Fever
Hepatitis A	Typhus	Schistosomiasis
Hepatitis E		
Polio		

Source: Adapted from Meade & Earickson, 2000

quiring water for the vector" (Meade & Earickson, 2000). Table 11.3 includes most of the major waterborne infectious diseases.

Reportedly, the incidence of many diseases appears to have stabilized above pre-Desert Storm levels; however, the actual magnitude and contributory causes of the increases remain unclear. In many areas, particularly northern and southern regions, the increases probably are more attributable to political choices by the Iraqi regime, including inequitable restoration of health services, rather than the effects of sanctions and the UN embargo (DIA-AFMIC, 2001).

Waterborne Diseases Spread or Contracted Through Intestinal or Urinary Tract

Sanitation is extremely poor throughout the country, including major urban areas. Local food and water sources are heavily contaminated with pathogenic bacteria, parasites, and viruses. The overall risk from both cholera and diarrheal diseases is elevated during the warmer months, May through September. Cholera is caused by ingestion of a causative agent, primarily in water contaminated with feces or vomitus from infected humans.

Hepatitis A and hepatitis E can cause prolonged illness in a smaller percentage. Hepatitis A is highly endemic and may pose a major health risk to non-indigenous persons. Most Iraqis contract hepatitis A virus infection during childhood. Hepatitis E has been reported, but endemic levels are unclear (DIA-AFMIC, 2001). There is no vaccine against hepatitis E, and immunoglobulin prepared in Europe or the United States does not provide protection. (Benenson, 1995). As for many other enteric infections, avoidance of contaminated food and water is the only effective protective measure.

Typhoid is also caused by ingestion of a causative agent in food and water contaminated by feces or urine from infected humans. The risk is elevated in crowded, populated areas with poor sanitation, especially during warmer months (Benenson, 1995).

Poliomyelitis is a viral infection occurring in areas where sanitation is poor. There were only 10 cases in Iraq in 1989, but this number has increased as sanitation declines (Table 11.2). The larger urban areas of Iraq have the most concentrated cases of poliomyelitis because it is a disease that can be contracted by proximity to hosts of the virus. The overall ratio of hosts per population is higher in rural areas, however, owing to decreased medical care availability.

Waterborne Diseases Spread by Arthropods

During the warmer months of May to November, the climate and ecological habitat support large populations of arthropod vectors, including mosquitoes, ticks, and sand flies (DIA-AFMIC 2001). Significant disease transmission is sustained countrywide, including urban areas. Serious diseases may not be recognized or reported due to the lack of surveillance and diagnostic capability.

Malaria is the major vector-borne risk in Iraq, capable of debilitating a high percentage of people for a week or more (Benenson, 1995). Principal risk areas include rural and urban locales in the northern provinces of Dahuk, Ninawa, Arbil, Sulaimaniya, and At Ta'mim, where transmission occurs from April through November, with a seasonal peak during July through August. Scattered foci probably occur in central and southern areas from the Tigris-Euphrates River basin to the border with Iran. Transmission occurs from mosquitoes year-round, initially peaking during April through May, followed by a secondary peak during August through September (based on data from Iran). In endemic areas, transmission primarily occurs in rural areas up to 1,500 m (4,875 ft.) elevation. This is because higher elevations are cooler in temperature and mosquitoes require warmer temperatures to survive. Baghdad and the immediate vicinity probably are low risk as a result of increased awareness and government funding of pesticide-based mosquito larvae eradication programs.

Malaria is endemic at increasing levels in the northern rural areas, associated with Kurdish rice farming and lack of insecticides during 1993 and 1994. The practice of rice farming encourages stagnant water, a prime breeding habitat for mosquitoes. Low-level focal transmission probably occurs in isolated locations in central and southern areas, with risk greater in southern areas adjacent to Iran. Nearly all indigenous cases are caused by a form of malaria called plasmodium vivax. Plasmodium vivax is generally not life-threatening except in the very young, the very old, and in patients with concurrent disease or immunodeficiency (Benenson, 1995). The distribution of plasmodium vivax is consistent with areas under 1,500 m in elevation, as mosquitoes are not adapted for cool temperatures.

Sand fly fever transmission occurs primarily from April through November, peaking in July through September, coinciding with vector activity. Foci may occur throughout the country, with elevated risk in village and peripheral urban areas, paralleling the distribution of sand fly vectors. Risk may be limited along the southwestern border with Saudi Arabia. Sand fly fever is caused by the bite of an infective sand fly. *Phlebotomus papatasi*, the primary vector, is most active between dusk and dawn, has a limited flight range, is peridomestic in its breeding habits, and readily enters human habitations to feed. Other vector-borne diseases that are less significant, but exist, include typhus, spotted fever, rickettsioses, and plague.

CONCLUSION

Medical geography plays a key role in understanding Iraq's culture and also in disease prevention. The analysis of medical geography requires a synthesis of information taken from nearly every subfield of geography. The health of the Iraqi people can be attributed, but not limited to, the subject matter contained in every chapter of this book. Iraqis need to make great strides in the near future if they are to overcome the nutritional deficiency and diseases that plague them.

12

Conclusion

Jon C. Malinowski

Any geographic analysis of a country is naturally a snapshot in time. Iraq is once again experiencing a period of change, as it has done repeatedly from the time it was Mesopotamia. But when time inevitably drops another layer of sand on this historic area, beneath the surface will be a cultural and physical pattern likely to resemble the current one. This is the strength of understanding the geography of a country: it offers some recognition of what is likely to change and what is likely to remain.

The material in this book represents our best attempt to provide an introduction to the complex geography of this important country. In doing so, we hope it will raise further questions and open avenues for additional research.

What we have not been able to do, however, is to fully analyze Iraq in the context of the larger region it shares with its important neighbors. Water disputes cause tensions with Syria and Turkey, as does the status of the Kurds. The well-being of Shiite populations in the south periodically revives animosities with Iran that were at the heart of the Iran-Iraq War. Kuwait still remembers Iraqi aggression in its territory, and Saudi Arabia balances stability in the region with fundamentalist groups within its borders who are sensitive to the preservation of important Islamic historical sites within Iraq.

The lesson of this book should be that Iraq is not a single entity. Within its borders are a variety of physical and human regions that differ in their areal extent. Although these regions are generalizations based on common features, such as language or climate, they help us take a first step toward a more thorough understanding.

Contributors

ABOUT THE EDITOR

Jon C. Malinowski, Ph.D. earned his doctorate in geography from the University of North Carolina at Chapel Hill in 1995. He also holds a Masters degree from UNC–Chapel Hill and a Bachelor of Science in Foreign Service from the Edmund A. Walsh School of Foreign Service at Georgetown University, where he graduated in 1991. An expert in human navigation, the geography of childhood, and the geography of Asia, he has published numerous academic articles and book chapters. He is also the coauthor of two trade books, *The Summer Camp Handbook* and *The Spirit of West Point: Celebrating 200 Years*. Currently, he is an associate professor of geography at the United States Military Academy, West Point, New York.

ABOUT THE CONTRIBUTORS

Peter A. Anderson, Ph.D. earned his doctorate from the University of Utah in 1994. He also holds Bachelors and Masters degrees from the State University of New York at Albany. He is currently an assistant professor of geography at the United States Military Academy, West Point, New York.

Dennis D. Cowher earned a Master of Science degree in geography from Penn State in 2001. He is currently an assistant professor at his alma mater, the United States Military Academy, West Point, New York.

James B. Dalton, Ph.D. earned his doctorate in geography from the University of Minnesota in 2001. He also holds advanced degrees from the Naval War College and Gannon University. His undergraduate degree was earned from Providence College in 1979. He is currently an assistant professor of geography at the United States Military Academy, West Point, New York.

Jeffrey S. W. Gloede is a 1992 graduate of the United States Military Academy and a 1998 graduate of the University of Missouri-Rolla, where he earned a Master of Science degree. He is currently on the faculty of the United States Military Academy, West Point, New York.

Brandon K. Herl holds a Master of Science degree from Colorado State University. He is currently an assistant professor of geography at the United States Military Academy, West Point, New York, where he graduated in 1990.

Contributors

Wendell C. King, P.E., Ph.D., earned his doctorate in Environmental Engineering from the University of Tennessee in 1988. He also holds degrees from Tennessee Technological University and the Naval War College. A professional engineer, he is currently professor and head of the Department of Geography and Environmental Engineering at the United States Military Academy, West Point, New York.

Albert A. Lahood earned a Master of Arts degree in geography from Syracuse University in 2001 and a Bachelor of Science degree from Salem State College in 1992. He is currently an assistant professor at the United States Military Academy, West Point, New York.

Andrew D. Lohman is a 1989 graduate of the United States Military Academy and a 1999 graduate of the University of South Carolina, where he earned a Master of Arts degree in geography. He served on the faculty at the United States Military Academy from 1999 to 2002.

Patrick E. Mangin earned a Master of Arts degree from the University of Minnesota in 2000. He is currently an assistant professor of geography at the United States Military Academy, West Point, New York, where he graduated with a Bachelor of Science degree in 1990.

Eugene J. Palka, Ph.D. earned his doctorate from the University of North Carolina at Chapel Hill in geography. He also holds a Master of Arts degree from Ohio University and a Bachelor of Science from the United States Military Academy, where he graduated in 1978. He is currently professor and deputy head of the Department of Geography and Environmental Engineering at the United States Military Academy, West Point, New York. He is the author or coauthor of several books.

Richard P. Pannell is a 1989 graduate of the United States Military Academy and a 1999 graduate of the University of Wisconsin at Madison, where he earned a Master of Science degree in Geography. He served on the faculty of the United States Military Academy from 1999 to 2002.

Matthew R. Sampson, holds a Master of Arts degree from the University of Kansas and a Master of Education degree from Drury College. He currently is an assistant professor of geography at the United States Military Academy, West Point, New York, where he graduated in 1991.

Wiley C. Thompson is a 1989 graduate of the United States Military Academy and a 1999 graduate of Oregon State University, where he earned a Master of Science degree. He served on the faculty of the United States Military Academy from 1999 to 2002.

Bibliography

Jeffery S. W. Gloede

Air Force Combat Climatology Center. 1995. Operational Climatic Data Summary for Iraq, Jan 1995. https://www2.afccc.af.mil/cgi-bin/index_mil.pl?aafccc_info/products.html (last accessed 7 Feb 2002).

——. 1996. Narratives for Iraq. https://www2.afccc.af.mil/cgi-bin/index_mil.pl?aafccc_ info/products.html (last accessed 7 Mar 2002).

——. 1997. Narratives for Iraq. https://www2.afccc.af.mil/cgi-bin/index_mil.pl?aafccc_info/products.html (last accessed 7 Mar 2002).

——. 2001. Köppen Climate Classification. Geographic Information System Data (GIS). https://www2.afccc.af.mil/cgi-bin/index_mil.pl?aafccc_info/products.html (last accessed 7 Mar 2002).

Altapedia Online 2002. Altapedia.com. http://www.altapedia.com/online/countries/iraq.com. (last accessed 5 March 2002).

Anderson, Ewan W. 2000. *The Middle East: Geography and Geopolitics*. New York: Routledge.

American University, 2002. http://www.american.edu/ted/ATATURK.HTM TED (last accessed 11 March 2002).

Animal Info. 1999. Animal Info-Iraq. http://www.animalinfo.org/country/iraq.htm. (last accessed 7 March 2002)

Arab.Net 2002. ArabNet-Iraq, Tour Guide. http://www.arab.net/iraq/tour/iraq_tour.html (last accessed 5 March 2002).

Army Environmental Policy Institute. 1995. "Health and Environmental Consequences of Depleted Uranium Use in the U.S. Army: Technical Report." http://www.aepi.army.mil/Library (last accessed 28 February 2002).

BBC News. 2000. "Turkish Dam Controversy." 22 January, 2000. http://news.bbc.co.uk/hi/english/world/europe/newsid_614000/614235.stm.

Beaumont, Peter, G. H. Blake, and J. M. Wagstaff. 1988. *The Middle East: A Geographical Study*. 2d Ed. New York: John Wiley.

Benenson, Abram S. Ed. 1995. *Control of Communicable Diseases. Army Field Manual 8–33*. 16th Ed. Washington, DC: American Public Health Association.

Britannica.com Inc. 2001. Various articles. In *Britannica 2001 Standard Edition CD-ROM*. Chicago, IL: Britannica.com Inc.

Central Intelligence Agency (CIA). 2001. *The World Factbook*. Washington, DC: Government Printing Office.

Bibliography

——. 2001. *The World Factbook*. Washington, DC: The Library of Congress, 2001, available online at: http://www.odci.gov/cia/publications/factbook/index. html.

——. 2001. *The World Factbook—Iraq*. Washington, DC: The Library of Congress, 2001, available online at: http://www.odci.gov/cia/publications/factbook/geos/iz.html (last accessed 3 April 2002).

——. 2002. *The World Factbook*. Washington, DC: The Library of Congress, 2001, available online at: http://www.cia.gov/cia/publications/factbook/index. html (last accessed 9 March 2002).

——. 1993. *Atlas of the Middle East*. Washington DC: U.S. Government Printing Office.

Ciment, James. 1996. *The Kurds: State and Minority in Turkey, Iraq and Iran*, New York: Facts on File.

Clark, Ramsey. 2000. "Letter to UN Security Council, 26 January 2000." http://www.iacenter.org/rc12600.htm (last accessed 2 March 2002).

Clausewitz, C. 1976. *On War*. Howard & Paret (Translators). Princeton, NJ: Princeton University Press.

CountryWatch 2001. *Country Review: Iraq*. http://www.countrywatch.com.

CountryWatch 2003. *Country Review: Iraq*. http://www.countrywatch.com.

Darvish, Tikva. 1987. "The Jewish Minority in Iraq: A Comparative Study of Economic Structure." *Jewish Social Studies*, Spring 87, Vol. 49 Issue 2.

Davidson, Roderic H. 1968. *Turkey*. Englewood Cliffs, NJ: Prentice Hall.

de Blij, H. J., and Peter O. Muller. 2001. *Geography: Realms, Regions, and Concepts*, 10th Ed., New York: John Wiley.

Defense Intelligence Agency (DIA). Armed Forces Medical Intelligence Center (AFMIC). 2001. *Medical Environmental Disease Intelligence and Countermeasures (MEDIC)*. CD-ROM.

Eickelman, Dale F., and James Piscatori. 1996. *Muslim Politics*. Princeton, NJ: Princeton University Press.

Encyclopedia Britannica Online 2001. Encyclopedia Britannica: Iraq. http://search.eb.com/bol/topic?idx_id608642&pm=1 (last accessed 22 February 2002).

Energy Information Administration 2001. *Iraq*. http://www.eia.doe.gov/ (last accessed 7 March 2002).

Ethnologue. 2000. *Ethnologue: Volume 1 Languages of the World*, 14th Ed. Barbara F. Grimes (Editor). Dallas, TX: Summer Institute of Linguistics.

Federal Research Division of the Library of Congress. 1988. Iraq Country/Area Handbook. From the Country/Area Handbook Program sponsored by the Department of the Army. http://lcweb2.loc.gov/frd/cs/iqtoc/html (last accessed 5 March 2002).

Food and Agriculture Organization 1997. *Special Report: FAO/WFP Food Supply and Nutrition Assessment Mission to Iraq*. http://www.fao.org/giews/english/alertes/srirq997.htm (last accessed 7 March 2002).

Geocities 2002. Iraq: Major Cities. http://www.geocities.com/iraqinfo/ (last accessed 5 March 2002).

Bibliography

Getis, A., J. Getis and J. D. Fellmann. 2001. *Introduction to Geography*. Boston: McGraw-Hill.

Gibson, McGuire. 1998. Nippur—Sacred City of Enlil Supreme God of Sumer and Akkad. (Originally appeared in Al-Rafidan, Vol. XIV, 1993). The Nippur Expedition, Oriental Institute, University of Chicago. http://www-oi.uchicago.edu/OI/PROJ/NIP/PUB93/NSC/NSC.html (last accessed 5 March 2002).

Goode's World Atlas, 20th Ed. 2000. Rand McNally.

Held, Colbert C. 2000. *Middle East Patterns: Places, People, and Politics*, 3rd Ed. Boulder: Westview Press.

Hiro, Dilip, 1996. *Dictionary of the Middle East*, New York: St. Martin's.

Hitti, Philip K. 1968. *Makers of Arab History*. New York: Harper & Row.

——. 1961. *The Near East in History*. Princeton, NJ: Van Nostrand.

Iraq Country Review, 2001–2002. CountryWatch.com. http://www.countrywatch.com/files/ (last accessed 22 February 2002).

"Iraq Land and Climate." 2002. http://www.geocities.com/iraqinfo/sum/land-climate.html (last accessed 26 February 2002).

Iraq Ministry of Health. 2001. "The Effect of Embargo on Health Status in Iraq." http://www.uruklink.net/health/epage1.htm (last accessed 28 February 2002).

——. 2001. Environment Improvement Activities. http://uruk.uruklink.net/health/epage2.htm (last accessed 2 March 2002).

Iraqioasis.com 2002. "History of Mesopotamia." http://www.iraqioasis.com/mesopotamia.html (last accessed 5 March 2002).

Izady, M.R. 1997. Environment and Ecology. http://www.kurdish.com.

Johnson, Larry. 1999. "A Nation Sagging Under the Weight of Sanctions." *Seattle Post-Intelligencer On Line*. http://seattlepi.nwsource.com/iraq/life1.shtml (last accessed 2 March 2002).

Johnston, R. J., Derek Gregory and David M. Smith. Eds. 1996. *The Dictionary of Human Geography*. Cambridge, MA: Blackwell.

Johnston, et. al. 1994. *The Dictionary of Human Geography*. 3d Edition. Cambridge, MA: Blackwell.

Kemp, Charles. 2000. "Iraqi Refugees." http://www3.baylor.edu/~Charles_Kemp/iraqi_refugees.htm (last accessed 2 March 2002).

Kriner, Stephanie. 2000. "Life in Iraq Deteriorates With U.N. Sanctions." The Red Cross Organization Web Site. http://www.redcross.org/news/archives/2000/2-7-00.html. (last accessed 2 March 2002).

Kelley, Marjorie. Ed. 1984. *Islam: The Religious and Political Life of a World Community*. New York: Praeger.

Kurbani, Agnes. 1995. *Political Dictionary of the Modern Middle East*. Lanham, MD: University Press of America.

"Kurdish Land." No date. http://www.ukin.org/land/land.htm.

"KURDISTAN: Land and Ecology." 1992. http://www1.920.telia.com/~u92002382/kurdistan.html.

Library of Congress. 1988. "Iraq—A Country Study." http://www.loc.gov (last accessed 26 February 2002).

McKee, LTC Kelly. 1991. "Diagnosis and Treatment of Diseases of Tactical Importance to US CENTCOM Forces." GulfLINK. http://www.gulflink.osd.mil/declassdocs/otsg/19961230/123096_sep96_decls23_0001.html (last accessed 4 March 2002).

McKnight, Tom L. and Darrell Hess. 2002. *Physical Geography: A Landscape Appreciation*, 7th Edition. Upper Saddle River, NJ: Prentice Hall.

Meade, Melinda S. and Robert J. Earickson. 2000. *Medical Geography*. 2nd Ed. New York: Guilford Press.

Metz, Helen Chapin. Ed. *Iraq: A Country Study*, Washington, DC: Headquarters, Department of the Army, US Government Printing Office, 1990.

Microsoft. 2002. "Iraq." *Microsoft Encarta Encyclopedia.* www.microsoft.com/encarta.

———. 2001. Microsoft Encarta. *Interactive World Atlas.*

National Oceanic and Atmospheric Administration. 2000. Climate Information Project. *2000 Archive of Climate-Weather Impacts.* https://www.cip.ogp.noaa.gov/Library/weekly/2000/08.29.00.html (last accessed 8 Mar 2002)

The National Gulf War Resource Center, Inc. 2001. "Area of Depleted Uranium Expenditure." http://www.ngwrc.org (last accessed 28 February 2002.)

Palka, Eugene J. Ed. 2001. *Afghanistan: A Regional Geography*. West Point, NY: Department of Geography & Environmental Engineering, United States Military Academy.

Palka, Eugene J. 2001. "Introduction to Urban Geography." *Physical Geography Study Guide*, 3d Edition. West Point, NY: United States Military Academy.

———. 2001. "Medical Geography." *Physical Geography Study Guide*, 3d Edition. West Point, NY: United States Military Academy.

Parker, Sybil. Ed. 1995. *World Geographical Encyclopedia*, Volume 3, Asia. New York: McGraw-Hill.

Polk, William R. 1991. *The Arab World Today*. Cambridge, MA: Harvard University Press.

Population Reference Bureau. 2003. *2002 World Population Data Sheet*. Washington, DC: Population Reference Bureau.

Relief Web. 2001. Map of Iraq Water Sanitation Programs. http://www.reliefweb.int/w/rwb.nsf (last accessed 2 March 2002).

Rubenstein, James M. 1996. *The Cultural Landscape: An Introduction to Human Geography*. Upper Saddle River, NJ: Prentice Hall.

Sela, Avraham. Ed. 1998. *Political Encyclopedia of the Middle East*. New York: Continuum Publishing Company.

Sun Tsu. 1963. *The Art of War*. S. B. Griffith (Translator). New York: Oxford University Press.

Tashiro, Akira. 2000. "Discounted Casualties: The Human Cost of Depleted Uranium." *The Chugoku Shimbun.* http://www.chugoku-np.co.jp/abom/uran/iraq_kids_e/index.html. (last accessed 28 February 2002).

Thompson, Wiley 2002. Country Map, "Location," in Jon Malinowski, ed., *Iraq: A Regional Geography.* West Point, NY: U.S. Military Academy Press, 2002.

United Nations Children's Fund (UNICEF). 1999. "Iraq Survey Shows Humanitarian Emergency." http://www.unicef.org/newsline/99pr29.htm (last accessed 28 February 2002).

United Nations. Office of the Iraq Programme (UNOIP). 2002. "Implementation of Oil-for-Food." http://www.un.org/Depts/oip/chron.html (last accessed 28 February 2002).

——. 2002. "Oil-for-Food. The Basic Facts." http://www.un.org/Depts/oip/latest/basfact_000610.html (last accessed 1 March 2002).

——. 2000. Population Policy Data Bank. "Iraq Population." www.un.org/esa/population/ publications/abortion/doc/iraq.doc (last accessed 2 March 2002).

——. 2002 Security Council. 1990. Resolution 661. S/RES/661. "Economic Sanctions Against Iraq." http://www.un.org//Depts/oip/scrs/scr661onu.htm (last accessed 28 February 2002).

——. 1991. Security Council Resolution 687 of 03 April 1991. http://www.un.org/Docs/scres/1991/687e.pdf (last accessed 7 March 2002).

——. 1995. Security Council Resolution 986 of 14 April 1995. http://www.un.org/Docs/scres/1995/9510988e.htm. (last accessed 9 March 2002).

United States Army. Center for Army Lessons Learned (CALL). 1990. "Winning in the Desert." http://www.army.mil/products/newsltrs/90-7/907toc.htm (last accessed 3 March 2002).

United States Census Bureau, 2002. http://www.census.gov/ipc/www/idbsum.html (last accessed 9 March 2002).

United States Committee for Refugees. 2002. http://www.refugees.org/world/countryrpt/mideast/iraq.htm (last accessed 9 March 2002).

United States Department of State. 2002. Background Notes: Iraq. Available online at: http://www.state.gov/r/pa/bgn/6804.htm

United States Geological Society. 2001. Earthshots: Satellite Images of Environmental Change: Iraq-Kuwait, 1972, 1990, 1991, 1997. http://www.cr.usgs.gov/earthshots/slow/Iraq/Iraq

Various Modern Iraq-related maps from the Perry-Castaneda Library Map Collection. The University of Texas Austin, UT Library Online. http://www.lib.utexas.edu/maps/iraq.html (last accessed 22 February 2002).

The World Gazetteer 2002. "The World-Gazetteer: Iraq" http://www.world-gazetteer.com/r/r_iq/htm (last accessed 5 March 2002).

World Health Organization. April 2001. "Depleted Uranium: Sources, Exposure and Health Effects." WHO: Geneva. http://www.who.int/environmental_information/radiation/Depluraniumintro.pdf (last accessed 9 March 2002).

World Resources Institute. 2001. Earth Trends: The Environmental Information Portal. http://earthtrends.wri.org/ (last accessed 11 March 2002).

Bibliography

World Wildlife Federation. Global 200: Blueprint for a Living Planet: Ecoregion 158, Mesopotamian delta and marshes. http://www.panda.org/resources/programmes/global200/pages/regions/region158.htm

——. No date. "The Wonder of Wetlands: The Threats." http://www.panda.org/resources/publications/water/wonder/page10-11.htm